Optical Sensing Technologies, Devices and Their Data Applications

Optical Sensing Technologies, Devices and Their Data Applications

Guest Editors

Yufei Chu

Hanqing Guo

Yuanxian Zhang

Abu Farzan Mitul

MDPI

Basel • Beijing • Wuhan • Barcelona • Belgrade • Novi Sad • Cluj • Manchester

Guest Editors
Yufei Chu
School of Marine and Atmospheric Sciences
Stony Brook University
Stony Brook
USA

Hanqing Guo
Department of Information and Computer Sciences
University of Hawai'i at Mānoa
Honolulu
USA

Yuanxian Zhang
School of Physics and Astronomy
Yunnan University
Kunming
China

Abu Farzan Mitul
Physical Chemistry and Applied Spectroscopy Group (C-PCS)
Los Alamos National Laboratory (LANL)
Los Alamos
USA

Editorial Office
MDPI AG
Grosspeteranlage 5
4052 Basel, Switzerland

This is a reprint of the Special Issue, published open access by the journal *Photonics* (ISSN 2304-6732), freely accessible at: https://www.mdpi.com/journal/photonics/special_issues/227308T3L1.

For citation purposes, cite each article independently as indicated on the article page online and as indicated below:

Lastname, A.A.; Lastname, B.B. Article Title. *Journal Name* **Year**, *Volume Number*, Page Range.

ISBN 978-3-7258-6964-0 (Hbk)
ISBN 978-3-7258-6965-7 (PDF)
https://doi.org/10.3390/books978-3-7258-6965-7

Contents

About the Editors

Yufei Chu

Yufei Chu is a Research Scientist at Stony Brook University with extensive expertise in Lidar Science and optical detection. He holds a Doctor of Engineering degree from the University of Science and Technology of China and completed postdoctoral research at the University of Colorado at Boulder and MSU. His primary research focuses on the remote sensing of the atmosphere using laser-based techniques, such as lidar, and leveraging machine learning for the analysis of optical datasets. He also specializes in the development of next-generation atmospheric sensing systems, fiber sensing, and applying laser ultrasonics for nondestructive evaluation (NDE). His work has appeared in high-impact journals including *Optics Express*, *Optics Letters*, the *Journal of Geophysical Research: Atmospheres*, *Remote Sensing*, and *Optics and Lasers in Engineering*.

Hanqing Guo

Hanqing Guo is an Assistant Professor in the Department of Electrical and Computer Engineering at the University of Hawaiʻi at Mānoa with extensive expertise in trustworthy artificial intelligence, audio and speech security, and privacy-preserving signal processing. He holds a Ph.D. degree from Michigan State University and has interdisciplinary training spanning computer science, signal processing, and communication systems. His primary research focuses on robust audio watermarking for generative AI, real-time voice authentication, and defenses against deepfake and voice-cloning attacks, with an emphasis on deployable, low-latency, and human-centered AI systems. He also specializes in low-power intelligent sensing and communication systems, secure speech and media technologies, and AI-driven signal processing for cybersecurity and media integrity. His work has appeared in top-tier venues including CCS, SenSys, and USENIX-affiliated conferences, and has received multiple recognitions such as a Best Paper Award and an Honorable Mention at leading security conferences.

Yuanxian Zhang

Yuanxian Zhang is a Professor in the School of Physics and Astronomy at Yunnan University, where he has been a faculty member since 2013. He earned his B.S. in Physics from Yunnan Nationalities University in 2006, followed by an M.S. in Optics and a Ph.D. in Theoretical Physics from Yunnan University in 2010 and 2013, respectively. To advance his research, he was a Visiting Scholar at the University of Texas at Arlington's Department of Electrical Engineering from 2018 to 2019, and subsequently at Nanyang Technological University's Department of Electrical and Electronic Engineering from 2024 to 2025. Dr. Zhang's work centers on the integration of photonics and microfluidics, with a focus on fiber optofluidic lasers and laser-based biochemical sensing. His recent innovations include the development of microfluidic-integrated ring resonator lasers and evanescent wave-excited waveguides, contributing significantly to high-sensitivity environmental and chemical monitoring, such as pH and nitrite detection. His research has been published in prominent journals including *Analytical Chemistry*, *Optics Express*, *The Analyst*, and the *Journal of Lightwave Technology*.

Abu Farzan Mitul

Abu Farzan Mitul is an American Bangladeshi academician and researcher. He completed his Ph.D. in Electrical Engineering from Michigan State University. He received his MSc in Optical Science and Engineering degree (majoring in Photonics) from the University of New Mexico. Prior to joining

Los Alamos National Laboratory (LANL) as a graduate student, he worked at the Center for Advanced Photovoltaics of South Dakota State University (SDSU), where he worked on the development of several advanced photovoltaic devices. He also worked at the National Energy Technology Laboratory (NETL) which is a U.S. national laboratory under the Department of Energy Office of Fossil Energy and Naval Medical Research Unit – San Antonio (NAMRU - SA) as an Oak Ridge Institute for Science and Education (ORISE) participant.

He previously worked as a Lecturer in the Department of Electrical and Electronic Engineering at Khulna University of Engineering and Technology (KUET) and taught courses as an Assistant Professor in the Department of Electrical & Computer Engineering at Saginaw Valley State University (SVSU), USA. At present, he is working as Research Scientist III at Leidos. His research focuses on fiber optic sensors and laser–material interactions.

photonics

MDPI

Editorial

Editorial for the Special Issue "Optical Sensing Technologies, Devices and Their Data Applications"

Yufei Chu

School of Marine and Atmospheric Sciences, Stony Brook University, Stony Brook, NY 11790, USA; yufei.chu@stonybrook.edu

1. Introduction

Optical sensing technologies represent a pivotal advancement in contemporary science and engineering, leveraging the unique properties of light to probe and analyze phenomena ranging from microscopic molecular interactions to large-scale atmospheric processes. As the Guest Editor of this Special Issue, I am pleased to introduce a comprehensive collection of 11 original research articles that push the boundaries of optical sensor design, device innovation, and data-centric applications. This Special Issue, now consolidated into a book format, highlights the interdisciplinary fusion of photonics, materials science, signal processing, and artificial intelligence, drawing from diverse contributions from researchers across Asia, Europe, and North America.

The overarching theme of this Special Issue, "Optical Sensing Technologies, Devices and Their Data Applications," was ways to tackle emerging challenges in areas such as environmental monitoring, biomedical imaging, industrial quality control, and atmospheric analysis. Techniques encompassing lidar, spectroscopy, fiber-optics, adaptive optics, and optofluidics provide exceptional sensitivity, non-contact operation, and resilience under demanding conditions. However, the full potential of these technologies is realized through sophisticated data handling, including machine learning algorithms and enhanced signal processing, which transform raw sensor outputs into meaningful intelligence. Manuscripts were solicited on topics like novel sensor architectures, hybrid opto-acoustic systems, AI-driven analytics, and waveguide innovations, with foci on gas detection, turbulence compensation, and biometric applications.

The 11 articles featured here form a cohesive narrative of theoretical insights, experimental breakthroughs, and practical deployments. They address critical issues such as noise reduction, polarization maintenance, and real-time data acquisition, while demonstrating applications from trace gas sensing to defect detection. As this Special Issue evolves into a dedicated book, it can serve as an essential resource for scholars and practitioners, fostering the integration of optical hardware with advanced computational frameworks. In the ensuing sections, I outline the articles, organized thematically to illuminate their synergistic contributions.

2. Advances in Optical Sensor Design and Gas Detection

Fundamental progress in optical sensing hinges on refined device configurations and methodologies that elevate detection precision and efficiency. This Special Issue showcases several works in this vein, emphasizing viscosity assessment, trace gas monitoring, and spectroscopic enhancements.

Contribution 1 presents a novel "viscosity–diffusion coupling" system employing a liquid-core cylindrical lens (LCL) for swift, concentration-dependent viscosity measurements. By imaging focal planes and numerically resolving Fick's second law alongside the Stokes–Einstein relation, the authors enable continuous viscosity profiling across 0–50% glycerol solutions at varying temperatures, aligning closely with established data and streamlining fluid analysis in industrial contexts.

In the area of atmospheric sensing, Contribution 2 advances a dual-wavelength lidar fusion approach at 355 nm and 532 nm for accurate atmospheric boundary layer (ABL) height determination. Weighting polarized signals by their signal-to-noise ratios and validating against radiosonde and ERA5 datasets, this method suppresses aerosol noise, delivering reliable ABL estimates in urban environments like Beijing, thereby supporting air quality and meteorological modeling.

Contribution 4 introduces a cost-efficient cavity ring-down spectroscopy (CRDS) system for ppb-level hydrogen sulfide (H2S) detection, featuring a digital locking circuit that boosts frequency precision 140-fold to 0.07 MHz uncertainty. With a 2 ppb detection threshold (3σ) and sub-ppb sensitivity per Allan variance, this analyzer fulfills industrial safety and environmental needs.

Complementing this theme, Contribution 7 enhances tunable diode laser absorption spectroscopy (TDLAS) for methane via Savitzky–Golay filtering in a 29.37 m path-length cell. Elevating the signal-to-noise ratio by 1.84 times, it achieves 0.53 ppm accuracy at 92 ppm levels, advancing high-precision monitoring for atmospheric and industrial uses.

These contributions collectively propel optical sensors toward greater sensitivity and applicability in gas and fluid detection.

3. Adaptive Optics and Turbulence Management

Turbulence poses a persistent obstacle in optical systems, impairing resolution and energy focus. This section highlights innovations in adaptive optics for simulation and correction, facilitating transitions from lab to real-world scenarios.

Contribution 3 describes a 500 Hz adaptive optical setup with dual bimorph deformable mirrors and tip-tilt correctors, integrated with a Shack–Hartmann sensor. It reconstructs turbulence aberrations, reducing them from 2.6 μm to 0.3 μm and minimizing focal spot jitter by 2–3 times, proving invaluable for astronomical and laser applications.

This focus on dynamic wavefront control bolsters data fidelity in turbulent settings, aligning with broader sensing reliability goals.

4. Optofluidics, Waveguides, and Fiber-Optic Innovations

The intersection of optics and microfluidics yields compact, versatile platforms for sensing and integration. The contributions to this theme explore waveguide arrays, polarized fibers, and tunable interferometers.

Contribution 5 unveils a broadband RGB fluorescent waveguide array using water-soluble CdSe/ZnS quantum dots (QDs) in an optofluidic chip. Analyzing propagation losses and optimizing concentrations, it extends to a 3×3 PDMS-based display, ideal for on-chip spectroscopy and biochemical sensing.

Contribution 10 models a fully polarized optical fiber with a prism head for plane wave compression, maintaining linear polarization via FDTD simulations. This design enhances the intensity for photonic integrated circuits and sensors, offering a simplified fabrication route.

Contribution 9 proposes an angle-tunable technique to optimize rear reflectance in Fabry–Pérot interferometers for fiber-optic ultrasound sensing. By modulating stress and adhesives to symmetrize transmittance curves, it minimizes errors and detects ultrasound signals at slope extrema, reducing costs for medical and structural monitoring.

These advancements underscore the potential of integrated optical structures in diverse sensing paradigms.

5. Data Acquisition and AI-Enhanced Processing Applications

The influx of optical data necessitates robust acquisition and AI tools for pattern extraction and decision-making. This Special Issue concludes with works integrating spectroscopy, deep learning, and vision for greenhouse gas monitoring, biometrics, and quality inspection.

Contribution 6 develops a dual-channel ADC method for interferogram acquisition in ground-based FTIR spectrometers, capturing high- and low-gain signals to boost the dynamic range. Applied to CO_2 retrieval, it doubles measurement accuracy, enhancing climate surveillance.

Contribution 8 introduces a dual-stream deep network with attention mechanisms for age estimation from transmission near-infrared dorsal hand vein images. On a 300-subject dataset, it leverages a CNN and residuals to focus on vein textures, outperforming baselines in biometric accuracy.

Finally, Contribution 11 optimizes YOLOv5 for cork disk quality detection in badminton manufacturing. Incorporating GAN-based defect synthesis, attention modules, and refined anchor strategies, it achieves a 95.1% F1 score and 178.5 FPS, surpassing originals in efficiency and precision for industrial vision.

These AI-infused methods not only amplify detection capabilities but also tackle practical challenges in data integrity and automation.

6. Conclusions

This Special Issue, now immortalized in book form, captures a transformative era in optical sensing, where device ingenuity converges with data analytics to address pressing global issues in environment, health, industry, and beyond. The 11 articles—from lidar fusion for atmospheric profiling to AI-optimized defect detection—exemplify strides in sensitivity, integration, and intelligence. They embody the collaborative ethos of our authors, whose varied perspectives have enriched this Special Issue.

As the Guest Editor, I express my profound thanks to all authors, reviewers, and the *Photonics* editorial staff for their dedication. This volume not only documents current achievements but also beckons future pursuits in quantum sensing, sustainable optics, and AI synergies. It is my hope that it ignites innovation among emerging photonics experts to unveil new horizons.

Funding: This editorial received no external funding.

Acknowledgments: The Guest Editor thanks all authors, co-editors (Hanqing Guo, Yuanxian Zhang, and Abu Farzan Mitul) for their invaluable support in developing this Special Issue and its book edition. Additionally, heartfelt thanks go to my wife, parents, and friends for their unwavering support and help throughout this process, as well as to my daughter for her adorable smile that brightened many days.

Conflicts of Interest: The author declares no conflicts of interest.

List of Contributions:

1. Wei, L.; Zhang, S.; Dai, B.; Zhang, D. Rapid Measurement of Concentration-Dependent Viscosity Based on the Imagery of Liquid-Core Cylindrical Lens. *Photonics* **2025**, *12*, 872. https://doi.org/10.3390/photonics12090872.
2. Fang, Z.; Li, S.; Yang, H.; Kuang, Z. A Dual-Wavelength Lidar Boundary Layer Height Detection Fusion Method and Case Analysis. *Photonics* **2025**, *12*, 741. https://doi.org/10.3390/photonics12080741.

3. Galaktionov, I.; Toporovsky, V. A Multi-Deformable-Mirror 500 Hz Adaptive Optical System for Atmospheric Turbulence Simulation, Real-Time Reconstruction, and Wavefront Correction Using Bimorph and Tip-Tilt Correctors. *Photonics* **2025**, *12*, 592. https://doi.org/10.3390/photonics12060592.
4. Xu, W.; Wang, X.; Zhao, L.; Zou, J.; Chen, B. Parts-per-Billion Detection of Hydrogen Sulfide via Cavity Ring-Down Spectroscopy. *Photonics* **2025**, *12*, 284. https://doi.org/10.3390/photonics12030284.
5. Li, D.; Chu, Y.; Xu, Q.; Liu, D.; Ruan, J.; Sun, H.; Li, J.; Guo, C.; Pu, X.; Zhang, Y. Quantum Dot Waveguide Array for Broadband Light Sources. *Photonics* **2025**, *12*, 212. https://doi.org/10.3390/photonics12030212.
6. Deng, Y.; Xu, L.; Jin, L.; Sun, Y.; Zhang, L.; Liu, J.; Liu, W. Research and Application of Interferogram Acquisition Method for Ground-Based Fourier-Transform Infrared Greenhouse Gas Spectrometer. *Photonics* **2025**, *12*, 38. https://doi.org/10.3390/photonics12010038.
7. Chen, S.; Tian, X.; Mu, T.; Yuan, J.; Cao, X.; Cheng, G. Enhancement of Methane Detection in Tunable Diode Laser Absorption Spectroscopy Using Savitzky–Golay Filtering. *Photonics* **2025**, *12*, 2. https://doi.org/10.3390/photonics12010002.
8. Shu, Z.; Xie, Z.; Zou, X. Dual-Stream Enhanced Deep Network for Transmission Near-Infrared Dorsal Hand Vein Age Estimation with Attention Mechanisms. *Photonics* **2025**, *11*, 1113. https://doi.org/10.3390/photonics11121113.
9. Chu, Y.; Alshammari, M.; Wang, X.; Han, M. Angle-Tunable Method for Optimizing Rear Reflectance in Fabry–Perot Interferometers and Its Application in Fiber-Optic Ultrasound Sensing. *Photonics* **2024**, *11*, 1100. https://doi.org/10.3390/photonics11121100.
10. Sun, W. Modeling a Fully Polarized Optical Fiber Suitable for Photonic Integrated Circuits or Sensors. *Photonics* **2024**, *11*, 961. https://doi.org/10.3390/photonics11100961.
11. Qu, L.; Chen, G.; Liu, K.; Zhang, X. A Study on the Improvement of YOLOv5 and the Quality Detection Method for Cork Discs. *Photonics* **2024**, *11*, 825. https://doi.org/10.3390/photonics11090825.

photonics

MDPI

Article

Rapid Measurement of Concentration-Dependent Viscosity Based on the Imagery of Liquid-Core Cylindrical Lens

Li Wei [1], Shuocong Zhang [1], Bo Dai [2,*] and Dawei Zhang [2]

1 College of Science, University of Shanghai for Science and Technology, Shanghai 200093, China; weilioptic@usst.edu.cn (L.W.); 242152146@st.usst.edu.cn (S.Z.)

2 Engineering Research Center of Optical Instrument and System, The Ministry of Education, Shanghai Key Laboratory of Modern Optical System, University of Shanghai for Science and Technology, Shanghai 200093, China; dwzhang@usst.edu.cn

* Correspondence: daibo@usst.edu.cn

Abstract: Viscosity is an inherent frictional characteristic of fluids that enables them to resist flow or deformation, thereby reflecting their flow resistance. It is significantly affected by concentration, but traditional viscosity measurements are limited to discrete concentrations, and multiple experiments are required for different concentrations, so the process is time-consuming. To overcome this limitation, this study presents a "viscosity–diffusion coupling" measurement system using a liquid-core cylindrical lens (LCL) as both the diffusion chamber and imaging element. It captures concentration profiles via focal plane imaging and solves Fick's second law and Stokes–Einstein relation numerically to determine the viscosity at varying concentrations. Experiments on the viscosity of glycerol solutions (0–50% mass fraction) at three temperatures were conducted and showed strong agreement with literature values. The method enables continuous viscosity measurement across varying concentrations within a single experiment, demonstrating reliability, accuracy, and stability in the rapid assessment of concentration-dependent viscosity.

Keywords: optical measurement; concentration-dependent viscosity; liquid-core cylindrical lens; finite difference method

1. Introduction

Liquid viscosity, as a crucial manifestation of the intermolecular forces within the liquid, stands as one of the key physical properties and technical indicators for characterizing the flow resistance of the liquid, and precise measurement of liquid viscosity holds great significance in a wide range of industrial sectors, including petrochemical, pharmaceutical, metallurgical, and food industries, as well as in scientific research. Consequently, liquid viscosity measurement techniques have consistently received substantial attention [1–5].

Traditional viscosity measurement methods obtain the liquid viscosity by characterizing their flow properties under specific conditions. The Falling Ball Method [6–8] determines viscosity by measuring the velocity of a ball descending in a liquid, where resistance is proportional to viscosity. It calculates viscosity through fall time measurement and is valued for its simple setup and efficiency. However, its precision is limited, and ensuring no turbulence during the ball's fall is crucial for accurate measurements. The rotational viscometer [9–11] calculates viscosity by measuring the resistance experienced by a rotating shaft in a liquid, which is proportional to viscosity. It requires determining rotational speed and torque. This method provides high precision and good repeatability. However, it has drawbacks including complex equipment, complicated procedures, and low efficiency.

The vibration method [12–14] determines viscosity by measuring the damping effect of a vibrating rod immersed in a liquid, where the damping is directly proportional to the viscosity. Viscosity is calculated based on amplitude and frequency data obtained during the vibration process. This method is particularly suitable for high-viscosity liquids and provides relatively high measurement precision. Nevertheless, its application is limited by the complexity of the equipment and the cumbersome operational procedures involved.

In addition, optical measurement methods are favored by researchers due to their non-contact nature and high precision. The optical tweezers technique [15–17] uses a focused laser to create an optical trap that captures micro- and nanoparticles. It measures viscosity by analyzing particle motion in the fluid, such as Brownian motion or forced vibration. However, the method requires precise optical components and nanoscale positioning stages. As a result, the system is complex and expensive. The interference method [18–20] calculates viscosity by analyzing light interference fringes in fluid films based on lubrication theory and optical interference principles, which enables non-contact, high-precision viscosity measurement, but it is also highly sensitive to environmental disturbances, and the required equipment is complex and expensive. In summary, the viscosity of liquids is significantly affected by factors such as temperature and concentration. The traditional viscosity measurement techniques discussed previously exhibit a common limitation: viscosity assessments are typically limited to discrete concentrations and struggle to capture the complex viscosity–concentration relationship. Multiple concentration measurements require separate experiments, leading to significant experimental effort.

To overcome the limitations previously identified, this study introduces a "viscosity–diffusion coupling" measurement system that employs a custom-designed LCL serving simultaneously as a liquid diffusion chamber and an imaging component. Utilizing focal plane imaging technology, the system captures experimental data reflecting the spatial concentration profile of the diffusing solution within the liquid core at discrete time points [21]. By applying the finite difference method to numerically solve Fick's second law, the numerical solutions are calibrated to closely match the experimental observations, enabling the determination of the diffusion coefficient $D(C)$ across varying concentrations. Subsequently, integrating the Stokes–Einstein relation allows for the calculation of solution viscosity at different concentrations. In the experiment, this approach effectively measures the viscosity of glycerol solutions within a concentration range of 0–50% by a single experiment at three temperatures (298.15 K, 303.15 K, 307.15 K), and the viscosity measurement deviation can reach 0.05 mPa·s.

2. Calculated Method of Liquid Viscosity

In general solutions, the interactions between solute molecules cause the diffusion coefficient D to depend significantly on the concentration C [22]. The relationship between the diffusion coefficient $D(C)$ and the concentration C can be expressed as

$$D(C) = D_0\left(1 + k_1 C + k_2 C^2 + k_3 C^3 + \cdots\right) \tag{1}$$

where k_1, k_2, and k_3 are undetermined coefficients, and D_0 represents the diffusion coefficient of an infinitely dilute solution, measured by transient image analysis method [21].

Fick's second law describes the law governing the temporal evolution of the concentration profile during the diffusion process [23], and the concentration profile $C(x,t)$ is governed by the one-dimensional diffusion equation (Equation (2)) [24]:

$$\frac{\partial C(x,t)}{\partial t} = \frac{\partial}{\partial x}\left[D(C)\frac{\partial C(x,t)}{\partial x}\right] \tag{2}$$

In the numerical calculation, the position of $x = 0$ is defined as the interface between the two diffusing solutions, with initial concentrations C_1 and C_2, respectively. The diffusion lengths of the two solutions are assumed to be long enough such that the concentrations at both interfaces always remain at the initial values C_1 and C_2. Given that the height of the LCL is L, the initial conditions and boundary conditions are as follows:

$$\begin{cases} t = 0 \\ L/2 > x > 0, C = C_1 \\ 0 > x > -L/2, C = C_2 \end{cases}, \begin{cases} \infty > t > 0 \\ x = L/2, C = C_1 \\ x = -L/2, C = C_2 \end{cases} \quad (3)$$

Generally, Equation (2) can be transformed as

$$\frac{\partial C(x,t)}{\partial t} = D(C)\frac{\partial^2 C(x,t)}{\partial x^2} + \frac{\partial D(C)}{\partial C}\left(\frac{\partial C(x,t)}{\partial x}\right)^2 \quad (4)$$

where Equation (4) is solved numerically using the implicit finite difference method [25]; the time step length of the diffusion process is Δt; the spatial domain L along the diffusion direction is discretized into N equally spaced nodes with a grid spacing Δx, respectively; the temporal points $t_i = i\Delta t$, where $i = 0, 1, 2, \ldots$; and spatial positions of the j-th node are defined as $x_j = j\Delta x$, where $j = 0, 1, 2, \ldots, N$. Equation (4) is rewritten as

$$\frac{C_j^{i+1} - C_j^i}{\Delta t} = \frac{(D_{j+1}^i - D_{j-1}^i)}{2\Delta x} \times \frac{(C_{j+1}^{i+1} - C_{j-1}^{i+1})}{2\Delta x} + D_j^i \times \frac{(C_{j-1}^{i+1} - 2C_j^{i+1} + C_{j+1}^{i+1})}{(\Delta x)^2} \quad (5)$$

where D_j^i is the difference form of the diffusion coefficient as a function of concentration, assuming $r_{j-1}^i = \Delta t \times D_{j-1}^i/\Delta x^2$, $r_j^i = \Delta t \times D_j^i/\Delta x^2$, and $r_{j+1}^i = \Delta t \times D_{j+1}^i/\Delta x^2$. Equation (5) is transformed as

$$C_j^{i+1} - C_j^i = \frac{r_{j+1}^i - r_{j-1}^i}{4} \times (C_{j+1}^{i+1} - C_{j-1}^{i+1}) + r_j^i \times (C_{j-1}^{i+1} - 2C_j^{i+1} + C_{j+1}^{i+1}), \quad (6)$$

Also, assuming $A_j^i = \left(r_{j+1}^i - r_{j-1}^i\right)/4$ and $B_j^i = r_j^i$, Equation (6) is simplified as

$$(A_j^i - B_j^i)C_{j-1}^{i+1} + (1 + 2B_j^i)C_j^{i+1} + (-A_j^i - B_j^i)C_{j+1}^{i+1} = C_j^i. \quad (7)$$

Subject to the initial and boundary conditions (Equation (3)), by employing the Thomas Algorithm [26], the numerical solution of the concentration profile $C^n(x,t)$ at time t can be obtained. Also, the experimental diffusion concentration profile $C^e(x,t)$ is obtained by the measurement platform shown in Figure 1. By changing k_1, k_2, and k_3, a series of $C^n(x,t)$s are calculated and the standard deviation σ_k can be calculated between the $C^e(x,t)$ and $C^n(x,t)$s based on Equation (8):

$$\sigma_k = \sqrt{\frac{\sum_0^N (C^n(x,t) - C^e(x,t))^2}{N}} \quad (8)$$

σ_k ($k = 2, \ldots, K$) is recalculated after selecting the undetermined coefficients [$(k_1)_n$, $(k_2)_n$, $(k_3)_n$]. Among the K groups of standard deviation values, the minimum value $\sigma_k = (\sigma_k)_{\min}$ is identified. When the standard deviation reaches its minimum, the undetermined coefficients are the optimal values $[k_1, k_2, k_3]_{\text{best}}$, and then these optimal coefficients are substituted into Equation (1) to obtain $D(C)$.

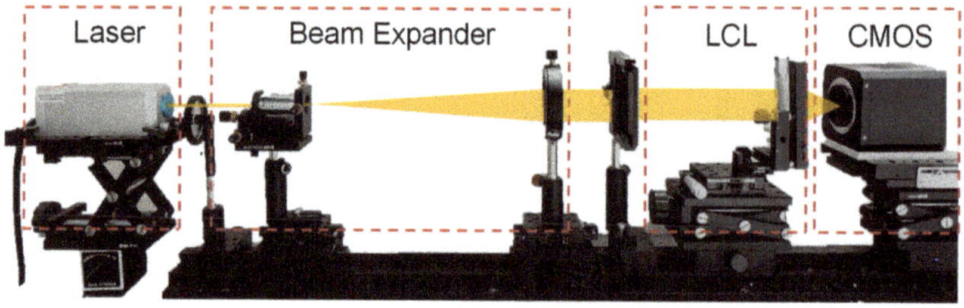

Figure 1. Experimental setup of the liquid viscosity measurement platform.

The concentration-dependent viscosity $\eta(C)$ is solely attributable to the diffusion coefficients, the hydrodynamic radius of the solute molecule, and absolute temperature. According to the Stokes–Einstein relation [27,28], the product of viscosity and diffusion coefficients is directly proportional to the absolute temperature:

$$\eta(C) = \frac{k_B T}{6\pi D(C) r_s} \tag{9}$$

where T is absolute temperature, k_B is the Boltzmann constant, and r_s is the hydrodynamic radius. The empirical relationship describing the viscosity η as an exponential function of concentration C can be derived by Equation (9).

3. Experimental Method for Measuring Liquid Viscosity

3.1. Experimental Setup

The liquid viscosity measurement platform, schematically illustrated in Figure 1, comprises the following optical components: a monochromatic laser (λ = 589 nm) is adopted as the working light source, to prevent damage to the image acquisition system caused by high power, and a neutral density filter is fixed behind the laser. A beam expander, composed of a microscope objective, a pinhole filter, and a collimating lens, and a slit are placed at the back. An LCL is formed by cementing two cylindrical lenses, its central liquid-core region functions as the diffusion chamber for the solution, and the spherical aberration is effectively eliminated through adjustments to the radius of curvature of the cylindrical lenses, with R_1= 32.0 mm, R_2 = 24.0 mm, R_3 = 34.7 mm, and R_4 = 79.5 mm, forming a high-resolution CMOS imaging system (6224 × 4168 pixels, with a pixel size of 3.76 μm).

3.2. Experimental Principle

The LCL was precisely filled with two liquids with distinct refractive indices (n_1 and n_2). Upon contact initiation, the liquids will diffuse with each other, resulting in the formation of a continuous refractive index gradient along the x-axis. When the CMOS sensor is positioned on the focal plane where the liquid with refractive index (n_c) can be sharply imaged, regions associated with other refractive indices exhibit defocused patterns, characterized by a distinct "beam-waist" diffusion image, as shown in Figure 2a. This relationship between the image width W and the localized refractive index arises from the fundamental imaging principles of LCL, as shown in Figure 2b, where the diffused width W and the size of the focal length f satisfy the following relationship [29]:

$$\frac{h}{f} = \frac{W/2}{|f - f_c|} \tag{10}$$

where the width of the collimated beam is h = 13 mm, limited by a variable-width rectangular slit; f_c is the focal length of the LCL with the refractive index of the liquid n_c; and f is the focal length of the LCL with the refractive index of the liquid n. Therefore, based on the width $W(x, t)$ of the diffusion image at a certain moment, the spatial profile $n(x,t)$ of the refractive index along the x direction is obtained.

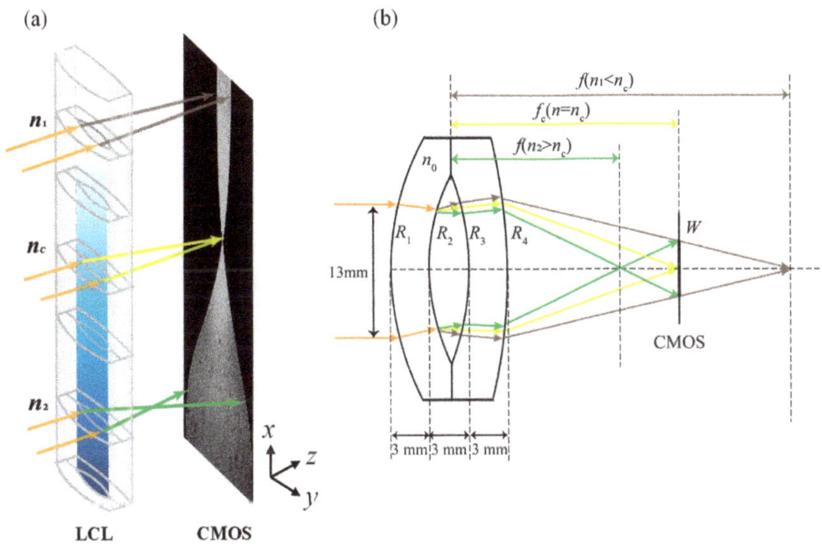

Figure 2. Schematic diagram of imaging principle. (**a**) Schematic diagram of imaging principle of LCL; (**b**) top view of imaging of LCL. The brown, yellow, and green lines represent the refractive rays with refractive indices of $n_1 < n_c$, n_c, and $n_2 > n_c$, which pass through the liquid core region.

4. Results and Discussion

4.1. Experimental Arrangement for Viscosity Measurement

The temperature of the experimental environment was maintained at 298.15 K. A glycerol solution with a mass fraction of 50% was injected using a microsyringe. The solution was then allowed to stand for 5–10 min to mitigate the effects of turbulent flow. Next, the upper liquid (deionized water) was slowly injected using a digital injection pump at a rate of 2.2 mL/min. The CMOS sensor was positioned where the collimated light could be sharply imaged with $n = n_c = 1.3443$. The diffusion process of the glycerol solution in the deionized water was recorded, with diffusion images captured every 5 min after the process began, and typical experimental diffusion images at different times are shown in Figure 3.

4.2. Experimental Concentration Profile from Diffusion Images

First, deionized water (refractive index n = 1.3327) was injected in the liquid core. The CMOS sensor was adjusted to the position x_1 where the refractive index can be sharply imaged at n = 1.3327, and the imaging system could be calibrated. Then, the position of the CMOS sensor was adjusted to the position x_2, where the liquid refractive index n_c = 1.3443 could be sharply imaged. Second, using a precision electronic balance (accuracy: 0.0001 g), eight sets of different concentrations of glycerol solutions ($\leq$50%) were prepared. Different concentrations of glycerol aqueous solutions were successively injected into the LCL, respectively, and their widths of images were recorded. The experimental results were fitted using the least squares method to establish a relationship between the image width

W (mm) and the refractive indices n, as shown in Figure 4a, and the fitting relation is as follows:

$$n = \begin{cases} -0.014W + 1.3454, \ n < n_c, \\ 0.013W + 1.3433, \ n > n_c. \end{cases} \quad (11)$$

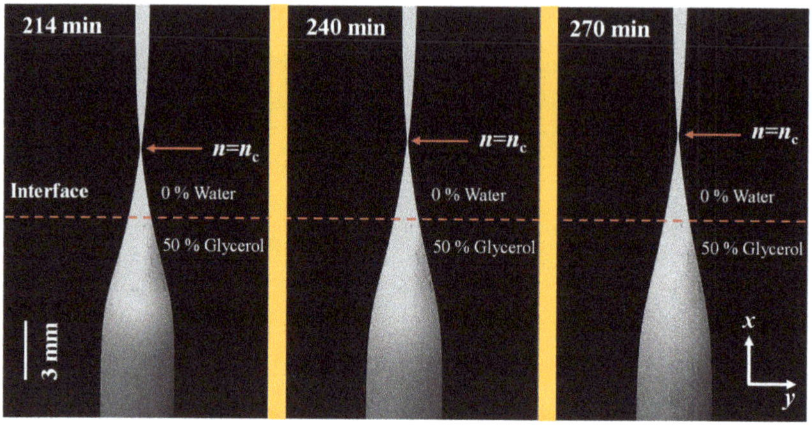

Figure 3. Typical experimental diffusion images at different diffusion times. The CMOS sensor was positioned where the collimated light could be sharply imaged with $n = n_c = 1.3443$. The red dashed line represents the interface between the two liquids at the initial moment $t = 0$.

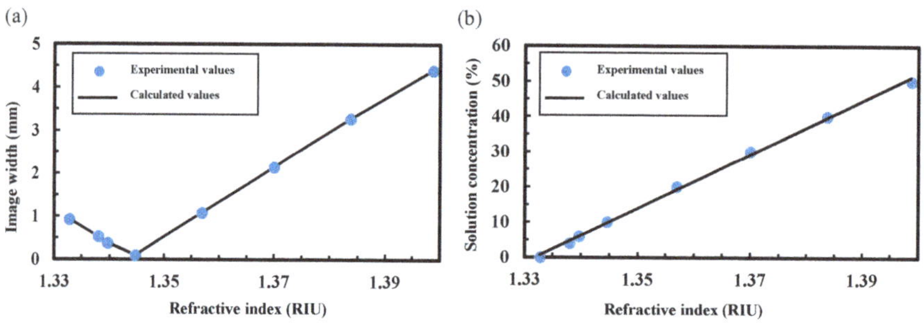

Figure 4. The relationship between image width, solution refractive index, and concentration is presented as follows: (**a**) image width corresponding to solutions with varying refractive indices, and (**b**) refractive index measurement results varying with different solution concentrations.

Furthermore, as illustrated in Figure 4b, the refractive indices of the glycerol solutions were measured using an Abbe refractometer. A linear regression analysis was conducted for the relationship between solution concentrations (0–50%) and refractive index, yielding a strong correlation coefficient (R = 0.9984). The resulting fitting equation is as follows:

$$C = 7.6452 \times n - 10.182 \quad (12)$$

4.3. Measurement Results of the Viscosity

Taking the experimental diffusion images at t = 214 min and 270 min as examples, binarization was performed to obtain the image width profile $W^e(x,t)$. Based on Equations (11) and (12), the refractive index profile $n^e(x,t)$ and diffusion concentration profile $C^e(x,t)$ along the diffusion direction were calculated, as represented by the blue dots in Figure 5.

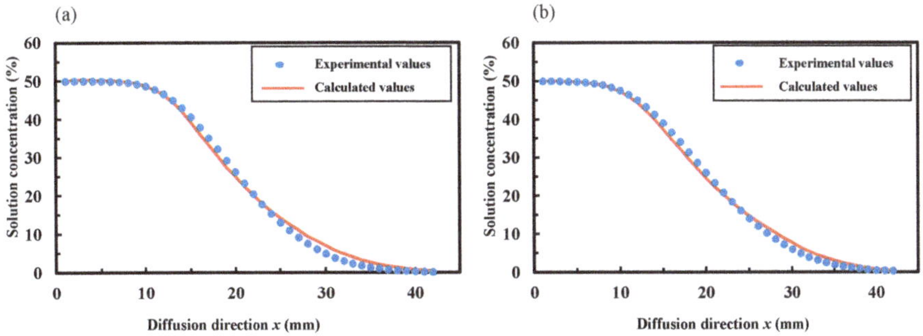

Figure 5. Temporal and spatial concentration profiles. The experimental concentration profile $C^e(x,t)$ is shown by the blue dots, whereas the calculated concentration profile $C^n(x,t)$ is illustrated by the red line, where (**a**) corresponds to t = 214 min and (**b**) corresponds to 270 min.

Subject to the initial and boundary conditions specified in Equation (3), a series of undetermined coefficients (k_1, k_2, k_3) were introduced to define the functional relationship $D(C)$. In the infinite dilute condition, $D(C)$ is independent of C, $D(C) = D_0$, Equation (2) has an analytical solution, the D_0 value has been measured by using the transient image analysis method introduced in reference [21], and $D_0 = 0.912 \times 10^{-5}$ cm^2/s. Then, using the finite difference method, the temporal and spatial concentration profile $C^n(x,t)$s at t = 270 min were calculated. The standard deviation σ_k between the experimental values $C^e(x, t)$ and the computed values $C^n(x, t)$ was calculated. By iteratively adjusting the undetermined coefficients $[(k_1)_n, (k_2)_n, (k_3)_n]$, the standard deviation σ_k (n = 2, 3, ..., K) was recalculated for each set of coefficients. Among the n groups of σ_k values, the minimum standard deviation $\sigma_k = (\sigma_k)_{\min}$ was identified. The corresponding optimal set of coefficients was determined as $[k_1 = (k_1)_{best} = -1.591, k_2 = (k_2)_{best} = -1.732, k_3 = (k_3)_{best} = 3.710]$. Based on this optimal coefficients, the best-fit polynomial of the concentration-dependent diffusion coefficient at t = 270 min was obtained as $D(C) = 0.912 \times 10^{-5}\,(1 - 1.591C - 1.732C^2 + 3.710C^3)$ cm^2/s, as shown in Figure 5, and the calculated concentration profile $C^n(x, t)$ showed good agreement with the experimental profile $C^e(x, t)$. In the experiment, eight diffusion images within the time range of 214 to 323 min were selected for analysis, so the average polynomial of ten times was $D(C) = 0.912 \times 10^{-5}\,(1 - 1.605C - 1.768C^2 + 3.644C^3)$ cm^2/s. Additionally, the aforementioned experiments were conducted at temperatures of 303 K and 307 K, and the results of these experiments can be summarized as follows, $D(C) = 1.020 \times 10^{-5}\,(1 - 2.432C + 2.890C^2 - 2.387C^3)$ cm^2/s and $D(C) = 1.120 \times 10^{-5}\,(1 - 2.372C + 2.742C^2 - 2.135C^3)$ cm^2/s, which are shown in Figure 6. According to the literature [27,28], there is no evidence indicating that glycerol forms stable large clusters at high concentrations. Consequently, its hydrodynamic radius approximately remains invariant with concentration, fulfilling the assumptions of Equation (9). The hydrodynamic radius was r_s = 0.27 nm, measured by the Dynamic light scattering method [30,31]. Using the experimentally obtained values of $D(C)$s, these data were substituted into Equation (9) to calculate the viscosity values $\eta(C)$ as a function of concentration. Empirical equations for the viscosity as a function of concentration (Equation (13)) were obtained through fitting at three temperatures. Multiple viscosity values at specific concentrations were selected, as presented in Table 1, and the measured values were compared with the corresponding values reported in the literature [32], revealing a close agreement between them.

$$\eta(C) = \begin{cases} 0.892 + 1.357C + 2.247C^2 + 16.740C^3 \text{ mpa} \cdot \text{s} & (298.15 \text{ K}) \\ 0.765 + 4.257C - 15.48C^2 + 38.332C^3 \text{ mpa} \cdot \text{s} & (303.15 \text{ K}) \\ 0.718 + 3.188C - 9.110C^2 + 24.982C^3 \text{ mpa} \cdot \text{s} & (307.15 \text{ K}) \end{cases} \quad (13)$$

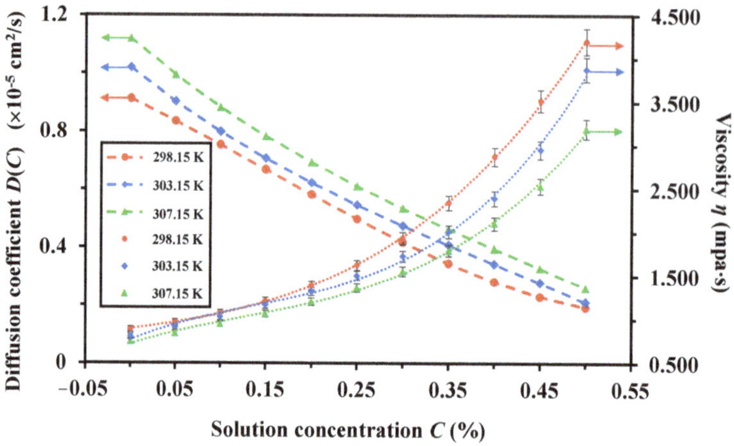

Figure 6. The diffusion coefficients and viscosities of glycerol solutions. The ambient temperature was set at 298.15 K, 303.15 K, and 307.15 K, and the error bar mainly stems from the measurement deviation of the hydrodynamic radius (±0.03 nm).

Table 1. Data of viscosities η (mPa·s) for aqueous glycerol solution from T = (298.15, 303.15, 307.15) K.

C/%	Method Using LCL			Method Using Capillary Tube [32]	
	298.15 K	303.15 K	307.15 K	298.15 K	303.15 K
0	0.886	0.806	0.744	0.897	0.803
5	0.968	0.910	0.837		
10	1.074	1.029	0.944	1.13	1.04
15	1.211	1.164	1.064		
20	1.391	1.321	1.203	1.49	1.32
25	1.626	1.506	1.364		
30	1.936	1.730	1.557	2.00	1.76
35	2.346	2.014	1.795		
40	2.875	2.395	2.105	2.84	2.48
45	3.520	2.953	2.535		
50	4.201	3.880	3.197	4.32	3.79

In this measurement methodology, the accuracy of experimentally determined concentration values directly influences the precision of viscosity measurements. The primary source of error in concentration determination arises from spherical aberration introduced by the LCL. Within the refractive index range of 1.3327 to 1.3972, the system-induced spherical aberration leads to an average imaging width deviation of δW = 28.23 μm, and this deviation corresponds to a concentration measurement error of δC = 0.30%, which in turn results in a calculated viscosity measurement deviation of $\delta\eta$ = 0.05 mPa·s.

5. Conclusions

In summary, the viscosity of liquids is significantly influenced by factors such as temperature and concentration, while traditional viscosity measurement methods are generally restricted to evaluating viscosity at discrete concentrations. Conducting viscosity measurements across multiple concentrations necessitates performing separate experiments for each concentration, thereby imposing a substantial experimental burden. To overcome the limitations previously identified, this study introduces a "viscosity–diffusion coupling" measurement system that employs a custom-designed LCL serving simultaneously as both a liquid diffusion chamber and an imaging component. Utilizing focal plane imaging technology, the system captures the experimental concentration profile of the diffusing solution. By applying the finite difference method to numerically solve Fick's second law of diffusion, numerical solutions of the concentration profile are calculated to closely match the experimental observations, enabling the determination of the diffusion coefficient $D(C)$ across varying concentrations. Subsequently, integrating the Stokes–Einstein relation allows for the calculation of solution viscosity at different concentrations. In this experiment, the viscosity of glycerol solutions within a concentration range of 0–50% at three temperatures (298.15 K, 303.15 K, 307.15 K) has been measured, with the results exhibiting strong agreement with the values documented in existing literature. This method requires only a single experiment to obtain viscosity values as a function of concentration, thereby offering a novel approach to viscosity measurement. Although the method requires preliminary determination of $n(C)$ and r_s, its benefits—in terms of experimental efficiency, reduced sample consumption, and enhanced data quality—underscore its value as an advanced and strategic technique.

Author Contributions: Methodology, Writing—Original Draft Preparation, Funding Acquisition, Conceptualization, L.W.; Formal Analysis, Data Curation, S.Z.; Writing—Review and Editing, Funding Acquisition, B.D.; Supervision, D.Z. All authors have read and agreed to the published version of the manuscript.

Funding: This research was funded by the National Natural Science Foundation of China, grant numbers [62205017, 62475157].

Institutional Review Board Statement: Not applicable.

Informed Consent Statement: Not applicable.

Data Availability Statement: The data that support the findings of this study are available from the corresponding author upon reasonable request.

Acknowledgments: The authors would like to express their sincere gratitude to the anonymous reviewers for their valuable feedback.

Conflicts of Interest: The authors declare no conflicts of interest.

Abbreviation

The following abbreviation is used in this manuscript:

LCL liquid-core cylindrical lens

References

1. Andrade, E.N.C. The viscosity of liquids. *Nature* **1930**, *125*, 309–310. [CrossRef]
2. Viswanath, D.S.; Ghosh, T.K.; Prasad, D.H.; Dutt, N.V.; Rani, K.Y. *Viscosity of Liquids: Theory, Estimation, Experiment, and Data*; Springer Science & Business Media: Berlin, Germany, 2007.
3. Atul, B. Review on viscosity measurement: Devices, methods and models. *J. Thermanal. Calorim.* **2023**, *148*, 6527–6543.
4. Leblanc, G.E.; Secco, R.A.; Kostic, M. Viscosity measurement. In *Mechanical Variables Measurement-Solid, Fluid, and Thermal*; CRC Press: Boca Raton, FL, USA, 2023.
5. Brooks, R.F.; Dinsdale, A.T.; Quested, P.N. The measurement of viscosity of alloys—A review of methods, data and models. *Meas. Sci. Technol.* **2005**, *16*, 354. [CrossRef]
6. Brizard, M.; Megharfi, M.; Mahé, E.; Verdier, C. Design of a high precision falling-ball viscometer. *Rev. Sci. Instrum.* **2005**, *76*, 025109. [CrossRef]
7. Ali, S.H.; Al-Zuky, A.A.D.; Al-Saleh, A.H.; Mohamad, H.J. Measure liquid viscosity by tracking falling ball Automatically depending on image processing algorithm. In *Journal of Physics: Conference Series*; IOP Publishing: Bristol, UK, 2019; Volume 1294, p. 022002.
8. Ballereau, P.; Truong, D.; Matias, A. Absolute falling ball viscometer, adapted to the low viscosities of liquids. *Int. J. Metrol. Qual. Eng.* **2016**, *7*, 305. [CrossRef]

9. Zhang, Y.; Huang, M.; Kan, Y.; Liu, L.; Dai, X.; Zheng, G.; Zhang, Z. Influencing factors of viscosity measurement by rotational method. *Polym. Test* **2018**, *70*, 144–150. [CrossRef]
10. Luo, H.; Huang, X.; Rongyan, T.; Ding, H.; Huang, J.; Wang, D.; Liu, Y.; Hong, Z. Advanced method for measuring asphalt viscosity: Rotational plate viscosity method and its application to asphalt construction temperature prediction. *Constr. Build. Mater.* **2021**, *301*, 124129. [CrossRef]
11. Zhou, Y.; Jin, Z.Y.; Yang, Z. Effects of surface tension and viscosity on the impact process of droplets onto a rotational surface. *Phys. Fluids* **2024**, *36*, 072110. [CrossRef]
12. Riesch, C.; Reichel, E.K.; Keplinger, F.; Jakoby, B.; Tenje, M. Characterizing vibrating cantilevers for liquid viscosity and density sensing. *J. Sensors* **2008**, *1*, 697062. [CrossRef]
13. Diogo, J.C.F.; Caetano, F.J.P.; Fareleira, J.M.N.A.; Wakeham, W.A. Viscosity measurements of three ionic liquids using the vibrating wire technique. *Fluid Phase Equilibr.* **2013**, *353*, 76–86. [CrossRef]
14. Ma, J.; Huang, X.; Bae, H.; Zheng, Y.; Liu, C.; Zhao, M.; Yu, M. Liquid viscosity measurement using a vibrating flexure hinged structure and a fiber-optic sensor. *IEEE Sens. J.* **2016**, *16*, 5249–5258. [CrossRef]
15. Statsenko, A.; Inami, W.; Kawata, Y. Measurement of viscosity of liquids using optical tweezers. *Opt. Commun.* **2017**, *402*, 9–13. [CrossRef]
16. Madsen, L.S.; Waleed, M.; Casacio, C.A.; Terrasson, A.; Stilgoe, A.B.; Taylor, M.A.; Bowen, W.P. Ultrafast viscosity measurement with ballistic optical tweezers. *Nat. Photonics* **2021**, *15*, 386–392. [CrossRef]
17. Zhang, Y.; Wu, X.; Wang, Y.; Zhu, S.; Gao, B.Z.; Yuan, X.C. Measurement of the microscopic viscosities of microfluids with a dynamic optical tweezers system. *Laser Phys.* **2014**, *24*, 065601. [CrossRef]
18. Verma, G.; Yadav, G.; Saraj, C.S.; Li, L.; Miljkovic, N.; Delville, J.P.; Li, W. A versatile interferometric technique for probing the thermophysical properties of complex fluids. *Light-Sci. Appl.* **2022**, *11*, 115. [CrossRef]
19. Guzmán, C.; Flyvbjerg, H.; Köszali, R.; Ecoffet, C.; Forró, L.; Jeney, S. In situ viscometry by optical trapping interferometry. *Appl. Phys. Lett.* **2008**, *93*, 184102. [CrossRef]
20. Ramírez-Miquet, E.E.; Perchoux, J.; Loubière, K.; Tronche, C.; Prat, L.; Sotolongo-Costa, O. Optical feedback interferometry for velocity measurement of parallel liquid-liquid flows in a microchannel. *Sensors* **2016**, *16*, 1233. [CrossRef]
21. Wei, L.; Meng, W.; Chen, Y.; He, B.; Zhou, Q.; Pu, X. Optical measurement of concentration-dependent diffusion coefficients of binary solution using an asymmetric liquid-core cylindrical lens. *Opt. Laser. Eng.* **2020**, *126*, 105867. [CrossRef]
22. Krause, A.L.; Gaffney, E.A.; Walker, B.J. Concentration-dependent domain evolution in reaction–diffusion systems. *Bull. Math. Biol.* **2023**, *85*, 14. [CrossRef]
23. Hashim, F.A.; Mostafa, R.R.; Hussien, A.G.; Mirjalili, S.; Sallam, K.M. Fick's Law Algorithm: A physical law-based algorithm for numerical optimization. *Knowl.-Based Syst.* **2023**, *260*, 110146. [CrossRef]
24. Crank, J. *The Mathematics of Diffusion*; Oxford University: Oxford, UK, 1975.
25. Lipnikov, K.; Manzini, G.; Shashkov, M. Mimetic finite difference method. *J. Comput. Phys.* **2014**, *257*, 1163–1227. [CrossRef]
26. Jena, S.R.; Senapati, A. Explicit and implicit numerical investigations of one-dimensional heat equation based on spline collocation and Thomas algorithm. *Soft Comput.* **2024**, *28*, 12227–12248. [CrossRef]
27. Elamin, K.; Swenson, J. Brownian motion of single glycerol molecules in an aqueous solution as studied by dynamic light scattering. *Phys. Rev. E* **2015**, *91*, 032306. [CrossRef]
28. Chen, B.; Sigmund, E.E.; Halperin, W.P. Stokes-Einstein relation in supercooled aqueous solutions of glycerol. *Phys. Rev. Lett.* **2006**, *96*, 145502. [CrossRef] [PubMed]
29. Sun, L.; Meng, W.; Pu, X. New method to measure liquid diffusivity by analyzing an instantaneous diffusion image. *Opt. Express* **2015**, *23*, 23155–23166. [CrossRef] [PubMed]
30. Pecora, R. Dynamic light scattering measurement of nanometer particles in liquids. *J. Nanopart. Res.* **2000**, *2*, 123–131. [CrossRef]
31. Linegar, K.L.; Adeniran, A.E.; Kostko, A.F.; Anisimov, M.A. Hydrodynamic radius of polyethylene glycol in solution obtained by dynamic light scattering. *Colloid J.* **2010**, *72*, 279–281. [CrossRef]
32. Shankar, P.N.; Manoj, K. Experimental determination of the kinematic viscosity of glycerol-water mixtures. Proceedings of the Royal Society of London. *Ser. A Math. Phys. Sci.* **1994**, *444*, 573–581.

photonics

MDPI

Article

A Dual-Wavelength Lidar Boundary Layer Height Detection Fusion Method and Case Analysis

Zhiyuan Fang [1,2], Shu Li [3,*], Hao Yang [2,4,*] and Zhiqiang Kuang [2,5]

1 School of Mechatronics and Energy Engineering, NingboTech University, Ningbo 315100, China; fangzhy@mail.ustc.edu.cn
2 University of Science and Technology of China, Hefei 230026, China; zqkuang@mail.ustc.edu.cn
3 Ningbo Langda Technology Co., Ltd., Ningbo 315100, China
4 School of Advanced Manufacturing Engineering, Hefei University, Hefei 230601, China
5 Hefei CAS GBo-Qua. Science and Technology Co., Hefei 230008, China
* Correspondence: shuli880818@nbt.edu.cn (S.L.); yh9599@mail.ustc.edu.cn (H.Y.)

Abstract: Accurate detection of the atmospheric boundary layer (ABL) is important for weather forecasting, urban air quality monitoring, and agricultural and ecological protection. In this study, we propose a new method for enhancing ABL height detection accuracy by integrating multi-channel polarized lidar signals at 355 nm and 532 nm wavelengths. Radiosonde observations and ERA5 reanalysis are used to validate the lidar-derived results. By calculating the gradients of signals of different wavelengths and weighted fusion, the position of the top of the boundary layer is identified, and corresponding weights are assigned to signals of different wavelengths according to the signal-to-noise ratio of the signals to obtain a more accurate atmospheric boundary layer height. This method can effectively mitigate the influence of noise and provides more stable and accurate ABL height estimates, particularly under complex aerosol conditions. Three case studies of ABL height detection over the Beijing region demonstrate the effectiveness and reliability of the proposed method. The fused ABLHs were found to be consistent with the sounding data and ERA5. This research offers a robust approach to enhancing ABL height detection and provides valuable data support for meteorological studies, pollution monitoring, and environmental protection.

Keywords: atmospheric boundary layer; lidar; ERA5

1. Introduction

The atmospheric boundary layer (ABL) typically extends from the surface up to approximately 1–2 km in altitude under average conditions, although its height can vary from several hundred meters to over 3 km depending on local meteorology and time of day [1–3]. The ABL plays a crucial role in the exchange of energy and matter between the Earth's surface and the free atmosphere [4–6]. Meteorological conditions are affected by surface characteristics such as topography, vegetation, and urban development, as well as by external factors, including solar radiation, ground thermal properties, and wind dynamics. These influences often manifest as significant gradient changes in elements such as temperature, humidity, and wind speed [7,8]. Based on the thermal stability of the lower atmosphere, the ABL can be classified into three types: the stable boundary layer, the neutral boundary layer, and the convective boundary layer [1,9]. There are various observation methods for the atmospheric boundary layer height (ABLH), mainly including traditional means such as radiosonde, satellite remote sensing, lidar detection, and machine learning methods [10–12]. Radiosonde measurements analyze the thermodynamic structure of the atmosphere by vertically profiling temperature, humidity, and wind speed, although their spatial and temporal coverage is limited, typically providing only two observations per day [13,14]. Satellite remote sensing technology monitors the ABL by using radiation data of different bands (such as infrared, microwave, visible light, etc.), and can provide multi-dimensional information such as temperature, humidity, aerosols, clouds, and wind fields [15–17]. With broad spatial coverage and high temporal resolution, it serves as a vital tool for weather forecasting, climate change research, and air quality monitoring. Nevertheless, satellite remote sensing faces challenges related to spatial resolution and atmospheric interference, and its data processing and interpretation still require further improvement [18,19]. Numerical simulation methods predict the behavior of the ABL by modeling atmospheric physical processes through computer algorithms, combined with observational data. They are well-suited for large-scale, continuous forecasting and are particularly advantageous under complex climatic conditions or extreme weather events [20–22]. However, the accuracy of numerical simulations depends on the quality of input data and the underlying model assumptions, and their high computational demands pose limitations for rapid-response applications [13,17,18].

The lidar detection method employed in this study represents a key application within the field of photonics. As the science of light generation, propagation, detection, and interaction, photonics provides both the theoretical foundation and technological support necessary for the high-precision observation of the ABL. Lidar, by emitting laser pulses and analyzing the returned signals, enables precise detection of airflow, turbulence, wind speed variations, and water vapor distribution within the ABL, thereby providing real-time dynamic information [23–25]. This capability makes it particularly effective under complex weather conditions. Among various types, aerosol lidar is the most commonly used method for detecting the atmospheric boundary layer height. The underlying principle is that aerosol particles are primarily concentrated within the ABL, with their concentrations significantly higher than those in the free atmosphere. In the transitional region between the ABL and the free atmosphere, the entrainment process causes the upper dry and clean air to descend into the ABLH [8]. The gradient method determines the ABLH by identifying the extrema in the aerosol concentration gradients of lidar return signals. Variants of this approach include the first-order gradient method, inflection point method, and logarithmic gradient method [26]. The threshold method defines the ABLH as the altitude where the lidar signal or its gradient exceeds a predefined threshold, though this typically yields relatively coarse results. The CWT method detects the ABLH by comparing the similarity between the lidar signal and wavelet functions, such as the Haar or Mexican Hat wavelets [27]. This technique is widely applied in the ABLH detection due to its ability to suppress noise and its suitability for processing complex aerosol structures and non-stationary signals. Despite its relatively high computational complexity, the continuous wavelet transform method offers significant advantages in practical applications, particularly for large-scale data analysis and high-precision extraction of the ABLH [14,18]. In addition, more methods have been proposed to handle complex aerosol structures, such as multi-layer recognition methods and image processing methods, to analyze the height of ABLH by analyzing the extreme values of backscattered images [10,14,18]. However, in the existing literature, most of these methods focus solely on the backscatter signal at a single wavelength, with other lidar detection parameters rarely considered [1]. The volume linear depolarization ratio (VLDR) is a key parameter for distinguishing aerosol types, as it reflects the morphology of atmospheric particles. It is defined as the ratio of the perpendicular to the parallel component of the lidar return signal [26]. A lower VLDR indicates more regular or spherical particles, whereas a higher VLDR suggests more irregular or elongated shapes [28,29]. Determining particle shape is highly relevant when probing the structure and extent of the atmospheric boundary layer (ABL) because aerosol shape directly influences optical properties and atmospheric behavior. Spherical and regular-shaped particles, such as liquid droplets, tend to occur above the ABL, while irregular or mixed aerosols, like dust or pollution, are more common within the ABLH. As a result, the vertical profile of depolarization provides valuable information for identifying the ABL height and transitions [6,18]. Therefore, it is essential to incorporate polarization measurements at 532 nm along with backscatter signals at 355 nm and 1064 nm when probing the structure and extent of the ABL. In this study, a two-wavelength signal-to-noise ratio weighted fusion algorithm is proposed for retrieving atmospheric boundary layer heights in Beijing. The accuracy of the method is analyzed through the application of three lidar observations at Huairou Station from 6 to 11 November 2014, in comparison with co-located soundings and ERA5 reanalysis. The method fills the gap of the single-wavelength method, which suffers from a lack of stability during soundings.

2. Data and Methods

2.1. Observation Site

This study primarily used ground-based meteorological and lidar data from Huairou, Beijing. The Huairou Meteorological Station, located in the Huairou District of Beijing, serves as a key site for meteorological research, climate monitoring, and early disaster warning. It provides essential atmospheric parameters, including temperature, humidity, wind speed, pressure, and precipitation, supporting comprehensive environmental and boundary layer studies. The station is situated in a suburban area northeast of central Beijing, characterized by relatively low building density, and surrounded by mixed terrain, including gentle hills, farmland, and some scattered residential and industrial developments. Compared to urban core sites, Huairou experiences fewer anthropogenic emissions and less urban heat island influence, making it suitable for studying natural boundary layer processes with minimal direct urban disturbance. During the period from 6 November to 10 November 2014, measurements were conducted twice daily at 07:15 and 19:15 Beijing time, at 12 h intervals.

2.2. Instruments

2.2.1. Lidar

Dual-wavelength polarization lidar offers the capability to directly observe the vertical structure of the atmosphere. It not only provides high resolution in the lower troposphere but also enables effective identification of the ABL top. The ground-based lidar system used in this study was developed by the Key Laboratory of Atmospheric Optics, Anhui Institute of Optics and Fine Mechanics, Chinese Academy of Sciences. It is a multi-wavelength aerosol lidar installed at the Yanqi Lake Campus of the University of Chinese Academy of Sciences (40.41° N, 116.68° E). The lidar is capable of stable, continuous 24 h operation.

To analyze the vertical distribution of aerosols, lidar experiments were conducted from 6 November to 10 November 2014. In this study, to regulate the operating condition of the laser, the lidar system was set to pause for 4 min after every 15 min of continuous operation. Measurements were suspended during rain or snow events. Due to the overlap between the emitted laser pulse and the return signal, which affects the accuracy of lidar measurements in the near-ground region, the lidar signal analysis in this study begins at an altitude of 225 m above ground level [16]. The specific parameters of the lidar system are listed in Table 1, and the system configuration is illustrated in Figure 1. The lidar system includes three receiving channels: a 355 nm channel, a 532 nm channel for parallel polarization ($\|$), and a 532 nm channel for perpendicular polarization ($\perp$). These channels are respectively used to measure aerosol scattering properties, backscattered signal intensity, and depolarization ratio.

Table 1. Detailed information on the multi-wavelength meter scattering lidar.

Technical Indicator	Parameter
Laser type	Nd:YAG
Laser wavelength/nm	355, 532
Receive channel	355 nm, 532 nm ǀǀ, 532 nm ⊥
Pulse energy/mJ	50,90
Pulse width/ns	20
Repetition rate/Hz	20
Receiving Telescope/mm	400 diameter, Cassegrain
Vertical resolution/m	7.5

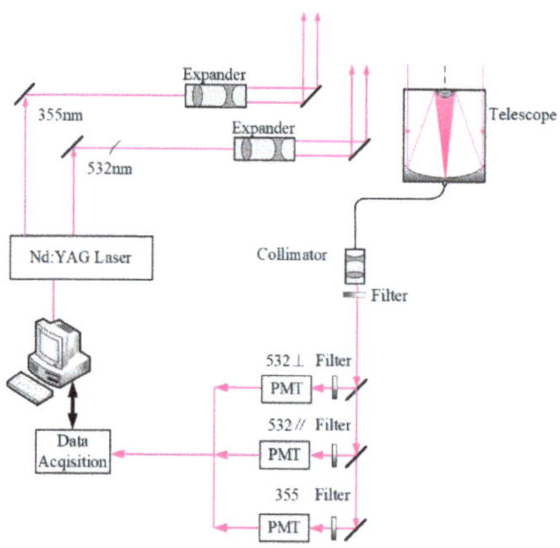

Figure 1. Schematic diagram of the lidar system.

2.2.2. Radiosonde

The radiosonde data used in this study were obtained from the Huairou region in Beijing. As part of Beijing's meteorological observation network, the Huairou sounding station plays a significant role in both meteorological monitoring and scientific research. The data collected were used to analyze the ABLH. In this study, the data were compiled and averaged with a vertical resolution of 10 m.

2.2.3. ERA5

The ERA5 dataset, provided by the European Centre for Medium-Range Weather Forecasts (ECMWF), is a global atmospheric reanalysis product designed to support climate research, weather forecasting, and a broad range of environmental sciences with high-precision historical meteorological data [30]. Released in 2017, ERA5 is a next-generation global reanalysis dataset covering atmospheric variables from 1950 to the present, with hourly temporal resolution. It offers a wide array of meteorological parameters, including temperature, humidity, wind speed, precipitation, and atmospheric pressure. ERA5 data are generated through the integration of numerical weather prediction models and observational data using advanced data assimilation techniques and complex numerical computations [31,32]. The dataset provides continuous meteorological fields and is widely applied in studies on climate change, weather forecasting, disaster monitoring, and model validation. Importantly, ERA5 not only delivers global spatiotemporal meteorological information but also allows users to download data at various spatial and temporal resolutions, making it highly valuable for both large-scale climate research and localized weather event analysis.

2.3. Different ABLH Algorithms

2.3.1. Traditional Lidar ABL Algorithm

The working principle of lidar involves emitting laser beams from a laser source and detecting the light scattered back by the atmosphere. Aerosol particles in the atmosphere play a key role in attenuating the laser signal due to their scattering effect, making them significant extinction factors in atmospheric laser transmission. In this study, the mixing layer refers to the convective portion of the ABL characterized by strong vertical turbulence and aerosol mixing. This study defines the mixing layer as the turbulent part of the ABL, where surface-driven convection leads to vigorous vertical mixing. It generally corresponds to the convective boundary layer during the daytime. Compared to the free atmosphere above, the mixing layer contains a higher concentration of aerosols, which enhances the scattering of the laser beam and contributes to stronger lidar backscatter signals. As a result, the ABLH can be readily identified by analyzing the gradient of the backscattered signal. The traditional lidar-ABLH consists of a gradient method, a threshold method, and a continuous wavelet variation method [26]. In the gradient method, the vertical gradient of the echo signal (or its logarithm) is first calculated, and the top of the ABL is located at the extreme value of the gradient plunge; in the threshold method, the signal intensity is compared with the empirical threshold of the free atmospheric background, and the first time that the echo is lower than this threshold is the ABLH, which is a simple method, but is more sensitive to the change in aerosol loading; the continuous wavelet transform (CWT) is used to convolve the logarithmic echo contour with the mother wavelet and search for mutant layers indicated by the pair of positive–negative extremes, and can automatically identify multi-layer structures, which is suitable for complex boundary layer scenarios. The CWT convolves the logarithmic echo profile with the mother wavelet to find the mutant layer labeled by the positive–negative extreme value pairs.

The lidar equation is typically expressed as follows [33]:

$$P(Z) = \frac{C}{Z^2}\beta(Z)\exp\left[-2\int_0^Z \alpha(Z')dZ'\right] \tag{1}$$

$$R(Z) = P(Z)Z^2 = C\beta(Z)\exp\left[-2\int_0^Z \alpha(Z')dZ'\right] \tag{2}$$

Here, C is the lidar constant, typically obtained through calibration. The backscatter coefficient $\beta(Z)$ represents the intensity of laser light scattered back toward the lidar per unit volume and per unit solid angle at a given height Z. It accounts for the combined contribution of atmospheric molecules (Rayleigh scattering) and aerosol particles (Mie scattering). In the lidar equation, $\beta(Z)$ serves as the primary source of the backscattered signal received by the detector, determining the strength of the return signal from the atmosphere. $\alpha(Z)$ represent the total backscatter coefficient and total extinction coefficient at height Z; the signal strength decays exponentially with distance during the laser propagation process. Therefore, $\alpha(Z)$ is the key parameter that determines the signal strength.

Since $\alpha(Z)$ and $\beta(Z)$ coexist and are coupled with each other in the lidar equation, it is impossible to directly separate the two based solely on single-channel echoes. Therefore, the following assumptions are often adopted in the inversion process:

$$S = \frac{\alpha(Z)}{\beta(Z)} \tag{3}$$

This formula reflects the ratio between the scattering and extinction of atmospheric components. Different types of aerosols have different Lidar ratios, and the general value is 50 [10,13].

2.3.2. New Fusion Algorithm

According to Mie scattering theory, the sensitivity of a laser to particles of a specific size can be explained using the size parameter, which is defined as [13,33]:

$$x = \frac{2\pi r}{\lambda} \tag{4}$$

where r is the particle radius and λ is the laser wavelength. According to Mie scattering theory, when the particle size is comparable to the wavelength of the incident light ($x \approx 1$), the scattering efficiency is significantly enhanced. Fine particulate matter with radii in the range of 100–200 nm falls within this sensitive range when irradiated with a laser at 355 nm. This enables the lidar system to effectively detect high aerosol concentrations in the lower atmosphere during polluted conditions, thereby facilitating accurate retrieval of the ABLH.

Polluted weather refers to conditions with high concentrations of aerosols and pollutants trapped within the atmospheric boundary layer due to weak winds, stable stratification, or temperature inversion. These conditions lead to strong near-surface lidar backscatter and can affect the detection of boundary layer height. High concentrations of aerosols, especially fine particulate matter such as PM2.5 and PM10, influence the dynamics of the ABL mainly through their interaction with solar radiation [10,14,16]. These aerosols scatter and absorb incoming solar radiation, reducing surface heating. This suppresses turbulent mixing in the ABL and lowers its height, thereby modifying its dynamic and thermodynamic structure. Under polluted weather conditions, the concentration of aerosols in the atmosphere increases significantly. This can directly affect the dynamic characteristics of the atmospheric boundary layer. Traditional single-wavelength lidar cannot accurately measure the height of the ABL, and multi-wavelength lidar provides an important means. The parallel and perpendicular components at 532 nm enable the characterization of scattering properties of various aerosol types, facilitating more accurate retrievals of particle type, concentration, and distribution. This capability is particularly valuable during dust transport events, where detailed vertical profiling is essential for identifying subtle variations in boundary layer structure. In urban aerosol detection, where particle radii are typically small, the 355 nm wavelength provides enhanced sensitivity to fine particles. Additionally, due to the presence of external aerosol sources such as dust in urban environments, the 532 nm channel complements the detection of coarser particles. This dual-wavelength approach improves responsiveness to high aerosol concentrations and haze conditions, strengthens boundary layer identification precision, and increases adaptability to various aerosol types and pollutant conditions. Therefore, this study proposes a novel method that combines the 355 nm and 532 nm parallel and polarization channels to improve the accuracy of ABLH detection.

First, the lidar signal undergoes range correction to eliminate attenuation caused by varying propagation distances. The signal is denoised and normalized to ensure consistency and comparability among different wavelength channels. Normalization maps the signal onto a unified scale, facilitating comparison and integration of different wavelengths. Next, the gradients of the 355 nm and 532 nm signals are computed to identify regions where the rate of change is minimal. These gradient minima typically correspond to the top of the ABL, characterized by sharp transitions in aerosol concentration. By calculating the gradient minima separately for the 355 nm and 532 nm channels, two independent estimates of the ABLH are obtained. Specifically, weights are assigned based on the SNR of each signal, with higher SNR channels contributing more to the final estimation. The fused ABLH, thus, represents an optimal integration of both signals. This approach not only effectively reduces noise and improves signal quality but also leverages the complementary information from multiple wavelengths to enhance the precision of boundary layer height detection. The fused ABLH is calculated as follows:

$$\mathrm{ABLH_{fused}} = \frac{\omega_{355} \cdot \mathrm{ABLH_{355}} + \omega_{532} \cdot \mathrm{ABLH_{532}}}{\omega_{355} + \omega_{532}} \quad (5)$$

$$\omega_{\lambda} = \frac{\mathrm{SNR}_{\lambda}}{\mathrm{SNR_{355}} + \mathrm{SNR_{532}}} \quad (6)$$

where ω_{λ} is the weight of different wavelengths, $\mathrm{ABLH_{fused}}$ is the height of the fused boundary layer, and SNR_{λ} is the signal-to-noise ratio. Here, the SNR for each channel is defined as the ratio of the mean signal intensity to its standard deviation. By calculating the gradient minima separately for the 355 nm and 532 nm channels, two independent estimates of the ABLH are obtained. Given the differing signal-to-noise ratios (SNRs) of each channel, a weighted fusion approach is applied to improve the reliability of the result. Specifically, weights are assigned based on the SNR of each signal, with higher SNR channels contributing more to the final estimation. The fused ABLH thus represents an optimal integration of both signals. This approach not only effectively reduces noise and improves signal quality but also leverages the complementary information from multiple wavelengths to enhance the precision of boundary layer height detection. The corresponding flowchart above is shown in Figure 2.

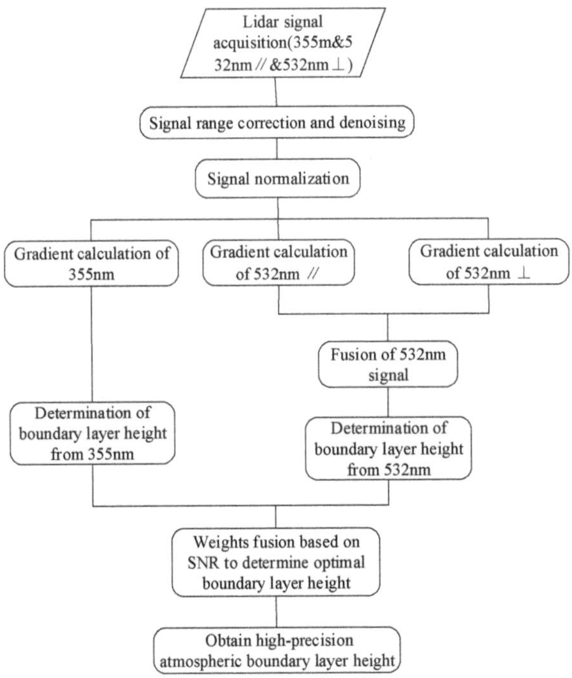

Figure 2. Flow chart of the new fusion algorithm to retrieve the ABLH.

Compared with traditional approaches—the gradient method, which is prone to signal-to-noise fluctuations and spurious extrema; the threshold method, which relies on empirical thresholds and is highly sensitive to aerosol loading; and the continuous wavelet transform

(CWT), which requires manual selection of mother-wavelet type and scale and involves substantial computation—the proposed dual-wavelength, SNR-weighted fusion algorithm combines the complementary gradient information from the 355 nm and 532 nm lidar channels. This dual-channel synergy suppresses noise and thin-cloud contamination, yielding stable ABLH estimates under both pristine and heavily polluted conditions. Because the scheme dispenses with external thresholds and wavelet parameters, it remains computationally light and suitable for real-time deployment while still automatically resolving multilayer boundary structures. Consequently, it offers a more reliable and generalizable determination of atmospheric boundary layer height in complex environments.

Figure 3 shows the lidar intensities and boundary layer heights at different time points calculated by the new method. In each pair of panels, the left panel shows the backscatter intensity profiles (black for 532 nm, green for 355 nm), and the right panel shows the corresponding vertical gradient profiles. Horizontal dashed lines indicate the ABLH estimates based on 532 nm (red), 355 nm (blue), and the final combined result (black). The signal intensities (a), (c), (e), and (g) on 4, 6, 21 and 22 November are selected to represent the signal intensities of different time periods, respectively, and (b), (d), (f), and (h) represent the corresponding gradient profiles after differentiation. As can be seen from the figure, the height of the boundary layer is higher during the day and lower at night.

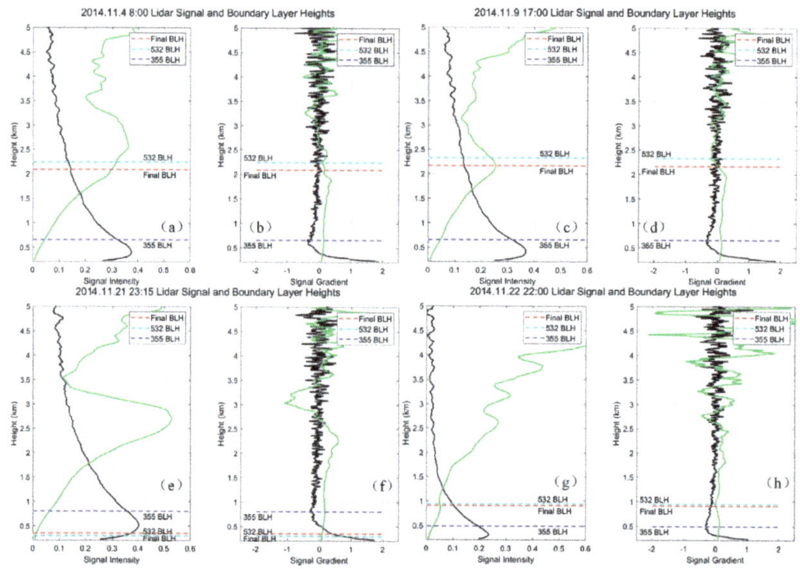

Figure 3. Lidar backscatter intensity and boundary layer height signals at different time points. (**a**) Signal intensity profile at 8:00; (**b**) Signal gradient profile at 8:00; (**c**) Signal intensity profile on at 17:00; (**d**) Signal gradient profile at 17:00; (**e**) Signal intensity profile at 23:15; (**f**) Signal gradient profile at 23:15; (**g**) Signal intensity profile at 22:00; (**h**) Signal gradient profile at 22:00.

2.3.3. Richardson Number Method

The Richardson number method is a widely used approach for analyzing the ABLH. This method determines the ABLH by calculating the Richardson number (Ri), which represents the ratio between atmospheric static stability and vertical wind shear. It provides insight into the dynamic and thermal structure of the atmosphere. The bulk Richardson number is typically calculated using the following formula [34]:

$$Ri(z) = \frac{\frac{g}{\theta_{vs}}(\theta_{vz} - \theta_{vs})(z - z_s)}{(u_z - u_s)^2 + (v_z - v_s)^2 + bu_*^2} \quad (7)$$

Here, z represents the height above the ground, and s represents surface-level quantities. θ_v is the virtual potential temperature, while u and v represent the horizontal wind components. u_* represents the surface friction velocity and b is a coefficient. In the calculation process, since the magnitudes of the wind components u and v are generally much larger than the surface friction velocity bu_*, the bu_* term is typically neglected.

According to Seidel et al. [35], the surface wind speed is set to zero in the Richardson number calculation. By integrating multiple meteorological factors such as temperature, humidity, pressure, and wind speed, the Richardson number method identifies the ABLH based on the thermodynamic characteristics of the atmosphere. One of the key advantages of this method is its comprehensive consideration of various atmospheric variables, which enables a relatively accurate estimation of ABLH. Comparative studies using critical Richardson number values of 0.20, 0.25, and 0.303 have shown that 0.25 provides the most reliable estimate for the ABLH [36,37]. This threshold reflects a balance between dynamic and thermal effects, making it particularly suitable for identifying the ABL top under varying atmospheric conditions. Therefore, in this study, a critical Ri value of 0.25 is adopted for retrieving the ABLH from the radiosonde. Figure 4 shows the ABLH obtained by calculating the temperature from the radiosonde data. It is acknowledged that the retrieved ABLH values, particularly in the afternoon profiles, may appear lower than expected. These results are dependent on the local meteorological conditions and the sensitivity of the Ri-based retrieval method. Due to the lack of direct validation data, the ABLH has not been verified at this stage.

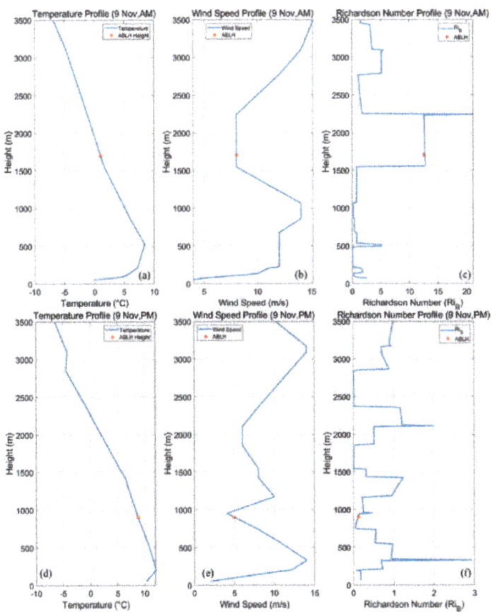

Figure 4. The ABLH determined from radiosonde data using the Richardson number method. (**a**,**d**) is the temperature profile, (**b**,**e**) is the wind speed profile, and (**c**,**f**) is the Richardson number profile, and the red point is the ABLH.

3. Results

Figure 5 shows the intensity of the lidar backscatter signal, with red color indicating stronger backscatter, illustrating the temporal variation in the ABL. Panels (a) and (b) display the vertical distributions of the depolarization ratio at 532 nm and the backscatter coefficient at 355 nm, respectively. The black and red dashed lines represent the ABLH obtained by the gradient method at 355 nm and 532 nm, respectively; the yellow curves represent the final fused ABLH results, and the pink dots correspond to the ABLHs derived from the radiosonde.

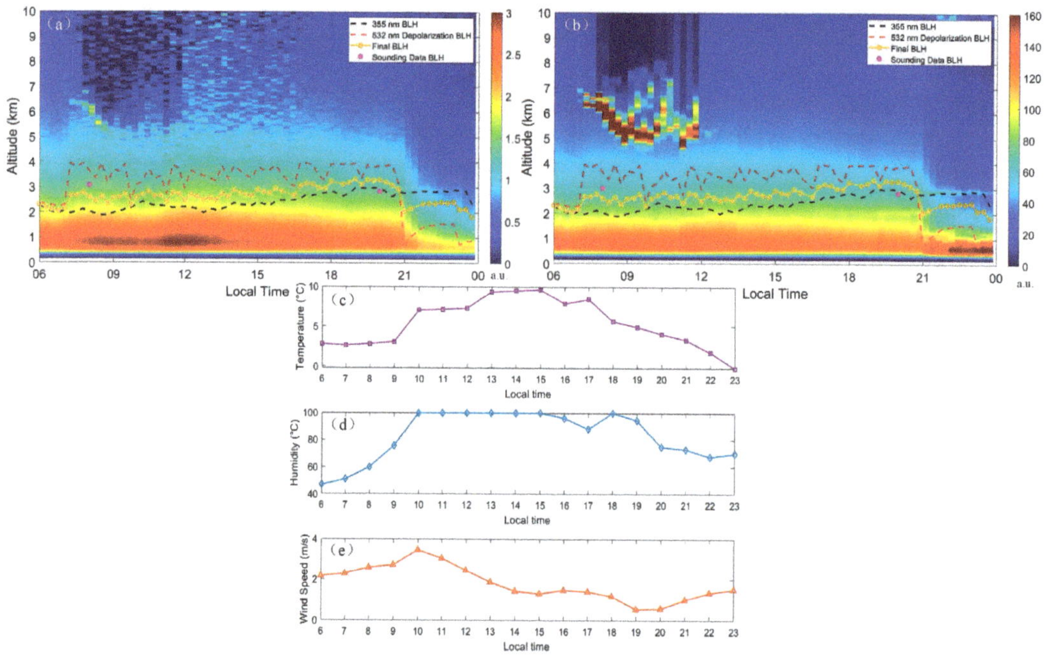

Figure 5. Variations in the ABLH and meteorological parameters from November 6 to November 7. (**a**) Depolarization ratio at 532 nm showing aerosol vertical distribution and ABLH changes. (**b**) Backscatter coefficient at 355 nm showing aerosol vertical distribution and ABLH changes. (**c**) Surface temperature variation over time. (**d**) Surface relative humidity variation over time. (**e**) Surface wind speed variation over time.

In the early morning, due to stable stratification and residual aerosol accumulation overnight, the ABLH remained around 1.5–2 km. Around midday, enhanced solar radiation induced surface heating and strong convective mixing, causing the ABLH to rapidly rise and peak at 2.5–3 km between 13:00 and 15:00. During this period, lidar occasionally detected thin cloud layers at altitudes of 6–7 km. Toward evening, the weakening of solar radiation and decay of turbulence led to a gradual decline in the ABLH. After sunset, rapid surface cooling and stabilized stratification caused the ABLH to drop to around 0.8 km, which was consistent with radiosonde observations.

The 532 nm retrieval generally overestimated the ABLH, while the 355 nm-derived values were lower before 21:00. Between 21:00 and 24:00, the 355 nm estimate exceeded that of 532 nm, indicating a higher concentration of fine aerosol particles in the atmosphere at that time.

Regarding surface meteorological conditions: at 06:00, temperature was as low as 4 °C, wind speed about 2 m/s, and relative humidity between 45 and 60%, resulting in a shallow

ABLH due to the lack of thermal and mechanical turbulence. After sunrise, the temperature rose quickly to 7–8 °C, reaching approximately 10 °C around 14:00, accompanied by an increase in wind speed to 3.5 m/s and a sharp rise in humidity to near saturation. These factors together intensified thermomechanical mixing and lifted the ABLH to its daytime peak. In the afternoon, both temperature and wind speed declined while humidity remained high, weakening surface fluxes and convection, leading to a gradual drop in the ABLH. By evening and into the night, temperature fell rapidly to 2–3 °C, wind became nearly calm, and humidity increased again, stabilizing the stratification and suppressing turbulence, causing the ABLH to collapse to approximately 0.8 km, consistent with radiosonde measurements. The observed abrupt variations in the ABLH—reaching up to approximately 1.5 km within a short time—are primarily caused by a combination of physical processes and lidar retrieval characteristics. During the daytime, intense solar heating of the surface leads to the rapid development of turbulence, allowing the mixed layer to rise from around 1 km to 2.5–3 km within 1–2 h.

Figure 6 illustrates the variations in the ABLH and meteorological parameters between 6:00 and 24:00 from 9 to 10 November 2014. In the early morning, the atmosphere was stably stratified, and residual aerosols supported a shallow boundary layer at approximately 1 km. After 08:00, surface heating due to solar radiation initiated thermal convection, and the ABLH rose and stabilized around 2.0–2.3 km. During this period, the 532 nm lidar channel detected aerosol tops reaching 3.5–4 km, indicating well-developed convective mixing, while the 355 nm-derived ABLH remained lower, underestimating the actual boundary layer height.

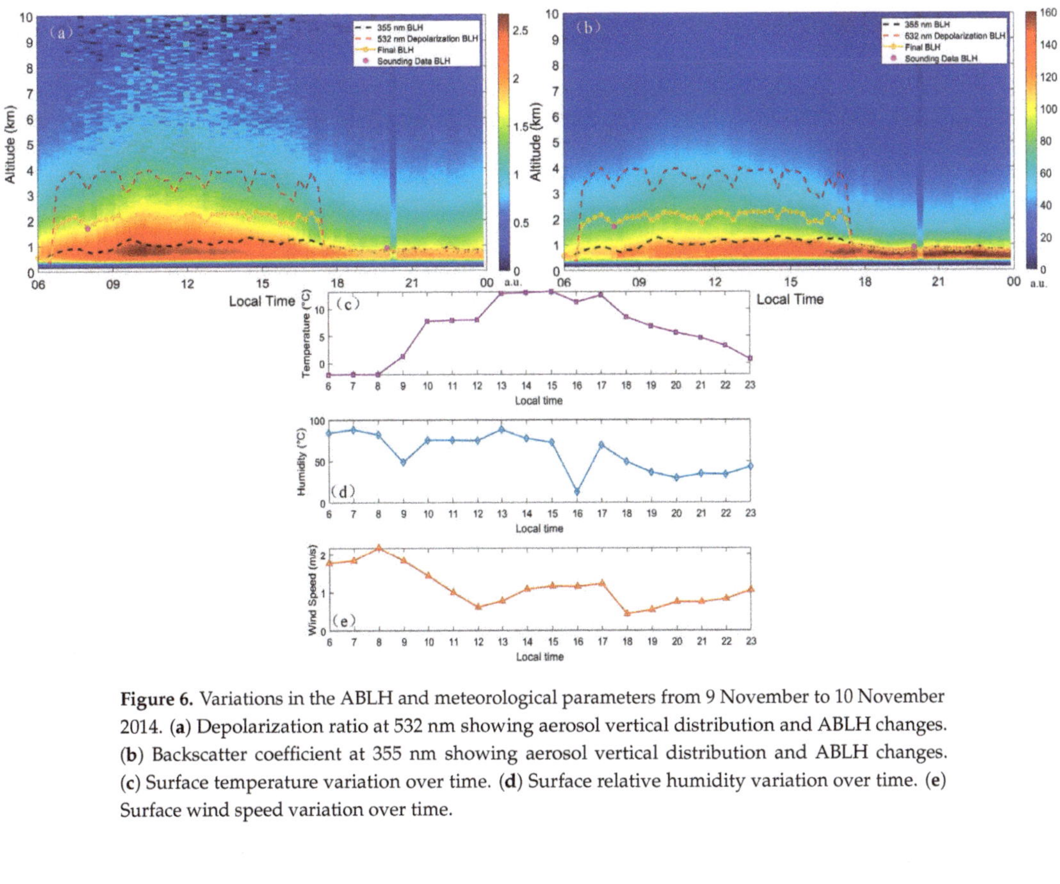

Figure 6. Variations in the ABLH and meteorological parameters from 9 November to 10 November 2014. (**a**) Depolarization ratio at 532 nm showing aerosol vertical distribution and ABLH changes. (**b**) Backscatter coefficient at 355 nm showing aerosol vertical distribution and ABLH changes. (**c**) Surface temperature variation over time. (**d**) Surface relative humidity variation over time. (**e**) Surface wind speed variation over time.

Around 16:00, a sharp drop in relative humidity combined with weak wind speed reduced surface heat flux, resulting in a declining trend in the ABLH. Following sunset, the rapid decay of turbulence caused the ABLH to collapse to about 0.5 km. After 18:00, a stable nocturnal layer formed, and the ABLH remained in the range of 0.6–0.9 km for an extended period. The ABLH retrieved by the fused algorithm closely matched those derived from radiosonde observations and proved to be more accurate than either the 532 nm or 355 nm channels alone.

Meteorological conditions support these findings: in the early morning, the temperature was only 0–1 °C, relative humidity ranged from 75 to 85%, and wind speed was around 2 m/s, maintaining atmospheric stability and suppressing the development of thermomechanical turbulence. After 09:00, the temperature rose rapidly to around 10 °C, wind speed initially increased to 2.5 m/s, then gradually weakened, and relative humidity generally stayed within 70–90%. Two abrupt drops in humidity (to 50% at 09:00 and to 10% at 16:00) indicate intrusions of dry, warm air aloft into the ABL. These conditions, combined with elevated temperatures and moderate wind, led to ABLH enhancement. However, in the late afternoon, the weakening of wind speed and the reduction in temperature and humidity gradients led to the simultaneous decline of both thermal and mechanical turbulence, resulting in a sharp drop in the ABLH. At night, temperatures fell to 1–3 °C, winds were nearly calm, and humidity increased to 40–50%, reinforcing stable stratification and trapping aerosols near the surface, thereby maintaining a low ABLH.

Figure 7 illustrates the evolution of the ABLH and meteorological parameters between 06:00 and 24:00 from 10 to 11 November 2014. During the early morning hours (06:00–08:00), a nocturnal inversion persisted, limiting vertical mixing, and residual aerosols kept the ABLH at approximately 1 km. As solar radiation intensified, convective development rapidly progressed between 09:00 and 15:00, lifting the ABLH and stabilizing it around 2.0–2.3 km. During this period, aerosol layers were mixed up to 3–3.5 km. Between 16:00 and 17:00, weakening surface heat flux and the intrusion of descending dry, warm air led to a marked decline in the ABLH, which fell below 1.5 km by sunset. After sunset, radiative surface cooling re-established a stable stratification, and the ABLH remained low, between 0.6 and 0.9 km, for the rest of the evening. The fused ABLH estimates more accurately captured the structure and evolution of the ABL than those derived individually from the 355 nm and 532 nm channels using the gradient method, showing close agreement with radiosonde observations.

From a meteorological perspective, the low ABLH during early morning was caused by the combined effects of a surface-based inversion and only moderate mechanical turbulence, which restricted mixing. After sunrise, rapid surface warming overcame the brief weakening of wind speed, and strong buoyancy-driven convection lifted the ABLH. Around midday, peak temperatures coupled with moderate wind speeds sustained efficient thermo-mechanical mixing, maintaining a high-level ABLH plateau. In the afternoon, the weakening of surface heat flux and the intrusion of dry, subsiding air caused a sharp drop in humidity and a reduction in mixing efficiency, initiating the decline in the ABLH. From evening into night, surface radiative cooling, near-calm winds, and rising humidity reestablished a stable stratification. As turbulence dissipated, the ABLH contracted sharply and remained at a low level.

Figure 8 compares the ABLH retrieved from the ERA5 reanalysis dataset with that obtained using the proposed fusion algorithm. To minimize the influence of outliers, extreme values were excluded during the analysis, and a linear fit was applied to the remaining data. The figure shows a clear linear trend: as the ERA5-derived ABLH increases, the ABLH estimated by the fusion algorithm also exhibits a strong upward trend. The correlation coefficient (R) of the fit is 0.79, indicating a moderate positive correlation between the two datasets. The root mean square error (RMSE) is 422.97, and the mean deviation (MD) is 290.17. The observed deviations can be attributed to both the limitations of the ERA5 dataset and the characteristics of the fusion algorithm. ERA5 is widely used for atmospheric reanalysis due to its comprehensive temporal and spatial coverage. It features a horizontal resolution of approximately 31 km. Vertically, it is structured with 137 model levels, with layer spacing near the surface ranging from tens to hundreds of meters. These spatial constraints hinder its ability to capture rapid variations in the local ABLH. In contrast, lidar measurements offer a much finer vertical resolution (e.g., 7.5 m), making them highly suitable for monitoring boundary layer structures. The coarser resolution of ERA5 data may thus fail to resolve localized meteorological processes, leading to significant discrepancies. The coarser resolution of ERA5 data may thus fail to resolve localized meteorological processes, leading to significant discrepancies. This is particularly the case if the fusion model is prone to overfitting or lacks robustness under specific boundary conditions. Meanwhile, ERA5, being based on a general circulation model, may not respond quickly to such localized changes, thereby further amplifying the deviation between datasets.

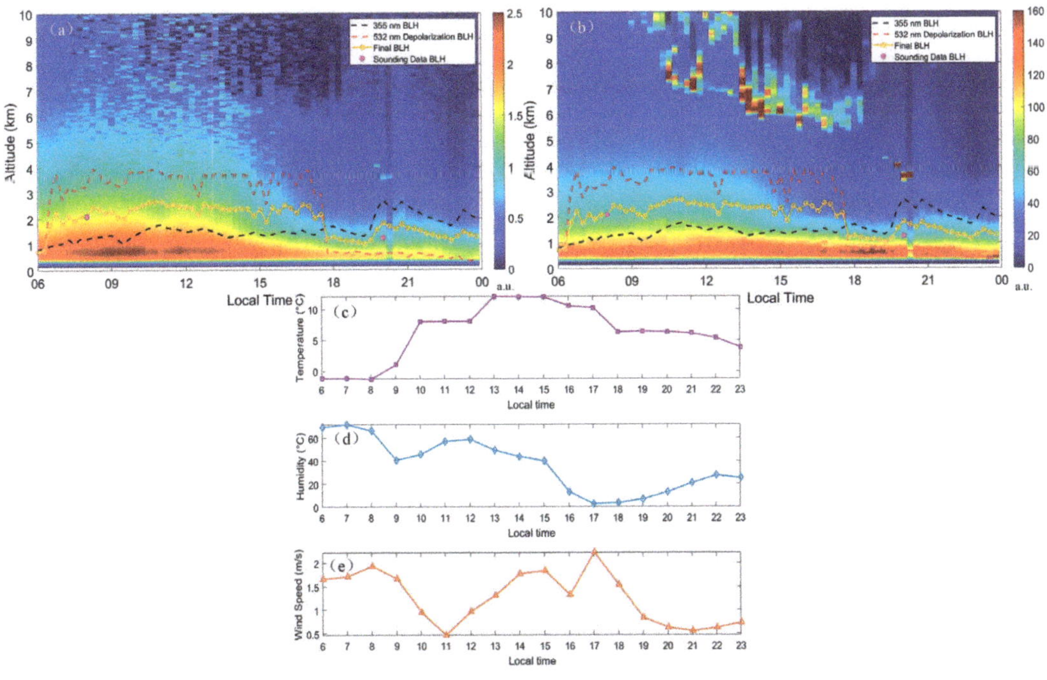

Figure 7. Variations in the ABLH and meteorological parameters from 10 November to 11 November 2014. (**a**) Depolarization ratio at 532 nm showing aerosol vertical distribution and ABLH changes. (**b**) Backscatter coefficient at 355 nm showing aerosol vertical distribution and ABLH changes. (**c**) Surface temperature variation over time. (**d**) Surface relative humidity variation over time. (**e**) Surface wind speed variation over time.

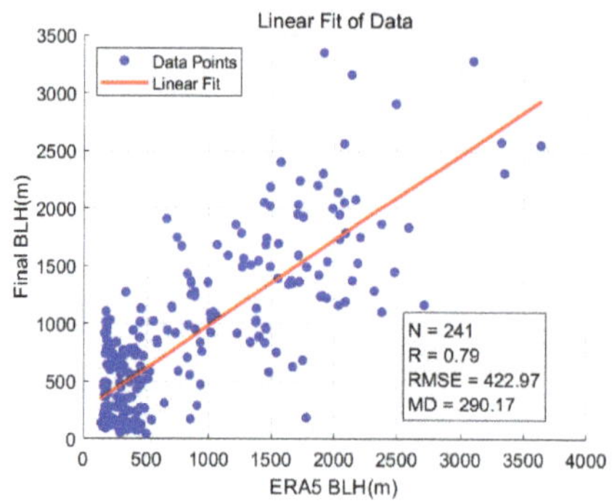

Figure 8. Scatter plot and linear fit between ERA5-derived boundary layer height and the ABLH predicted by the proposed fusion algorithm.

4. Conclusions

This study proposes a novel method for detecting the ABLH based on multi-channel lidar observations. By combining dual-wavelength lidar signals at 355 nm and 532 nm, the method significantly improves the accuracy and reliability of ABLH retrieval. Through a weighted fusion approach integrated with the gradient method, the algorithm enables precise identification of the ABL top, particularly under complex meteorological conditions and during episodes of severe air pollution.

The proposed method integrates lidar signals at 355 nm and 532 nm using a weighted fusion algorithm, yielding enhanced accuracy in boundary layer detection. Experimental results demonstrate that the fused ABLH estimates align well with radiosonde observations and ERA5 reanalysis data, showing strong adaptability and precision. Analysis of data from 6 November to 10 November 2014 reveals typical diurnal variations in the ABLH over Beijing. The ABLH is lowest in the early morning, rises rapidly with increasing solar radiation, and reaches its peak around midday. As solar heating weakens in the late afternoon, the ABLH gradually declines. The study also shows that the 532 nm channel tends to overestimate the ABLH, while the 355 nm channel underestimates it during certain periods—reflecting the influence of fine particulate matter on boundary layer structure. Compared with traditional single-wavelength lidar techniques, the dual-wavelength fusion method enables more accurate ABLH retrieval, especially in complex atmospheric environments. Experimental results across different time periods confirm that the fusion algorithm enhances detection accuracy and closely matches radiosonde-derived heights. When compared with ERA5 reanalysis data, the fused ABLH shows a correlation coefficient of 0.79, further validating the method's reliability.

Despite the effectiveness of the proposed multi-wavelength lidar fusion approach in improving ABLH detection under complex weather conditions, some limitations remain. For example, signal quality may be affected under cloudy conditions. Future work will focus on further refining the processing algorithm to enhance performance in the presence of complex aerosol structures. Incorporating additional meteorological variables and high-resolution datasets will also be important for validating the algorithm's robustness and general applicability.

Author Contributions: Conceptualization, Z.F. and H.Y.; methodology, Z.F.; software, S.L.; validation, Z.F. and Z.K.; formal analysis, H.Y.; writing—original draft preparation, Z.F.; writing—review and editing, H.Y.; supervision, S.L. and H.Y.; project administration, Z.F.; funding acquisition, Z.F. and S.L. All authors have read and agreed to the published version of the manuscript.

Funding: This research received no external funding.

Institutional Review Board Statement: Not applicable.

Informed Consent Statement: Not applicable.

Data Availability Statement: The data underlying the results presented in this paper are not publicly available at this time but may be obtained from the authors upon reasonable request.

Acknowledgments: The study is supported by Ningbo Tech University Talent Project (No. 20230419Z0106), the Key Research and Development Program of Ningbo (No. 2024Z257), The General Scientific Research Project of Zhejiang Education Department (Y202352444), the National Science Foundation of China (Grant No. 42405069) and the Fundamental Research Funds for the Central Universities lzujbky-2022-it26. The authors are grateful to ERA5 Analysis site who provided the meteorological data.

Conflicts of Interest: The authors declare that the research was conducted in the absence of any commercial or financial relationships.

References

1. Stull, R.B. *An Introduction to Boundary Layer Meteorology;* Springer: Dordrecht, The Netherlands, 1988.
2. Anand, M.; Pal, S. Exploring atmospheric boundary layer depth variability in frontal environments over an arid region. *Bound. Layer Meteorol.* **2023**, *186*, 251–285. [CrossRef]
3. Vivone, G.; D'Amico, G.; Summa, D.; Lolli, S.; Amodeo, A.; Bortoli, D.; Pappalardo, G. Atmospheric boundary layer height estimation from aerosol lidar: A new approach based on morphological image processing techniques. *Atmos. Chem. Phys.* **2021**, *21*, 4249–4265. [CrossRef]
4. Fang, Z.; Yang, H.; Cao, Y. Study of Persistent Pollution in Hefei during Winter revealed by Ground-based LiDAR and the CALIPSO Satellite. *Sustainability* **2021**, *13*, 875. [CrossRef]
5. Chu, Y.; Alshammari, M.; Wang, X.; Han, M. Angle-Tunable Method for Optimizing Rear Reflectance in Fabry–Perot Interferometers and Its Application in Fiber-Optic Ultrasound Sensing. *Photonics* **2024**, *11*, 1100. [CrossRef]
6. Su, T.; Li, J.; Li, C.; Xiang, P.; Lau, A.; Guo, J.; Yang, D.; Miao, Y. An intercomparison of long-term planetary boundary layer heights retrieved from CALIPSO, ground-based lidar, and radiosonde measurements over Hong Kong. *J. Geophys. Res. Atmos.* **2017**, *122*, 3929–3943. [CrossRef]
7. Melfi, S.H.; Spinhirne, J.D.; Chou, S.-H.; Palm, S.P. Lidar Observations of Vertically Organized Convection in the Planetary Boundary Layer over the Ocean. *J. Clim. Appl. Meteorol.* **1985**, *24*, 806–821. [CrossRef]
8. Banakh, V.A.; Smalikho, I.N.; Falits, A.V. Wind–Temperature Regime andWind Turbulence in a Stable Boundary Layer of the Atmosphere: Case Study. *Remote Sens.* **2020**, *12*, 955. [CrossRef]
9. Shi, L.; Zhu, A.; Huang, L.; Yaluk, E.; Gu, Y.; Wang, Y.; Wang, S.; Chan, A.; Li, L. Impact of the planetary boundary layer on air quality simulations over the Yangtze River Delta region, China. *Atmos. Environ.* **2021**, *263*, 118685. [CrossRef]
10. Li, C.; Li, J.; Dubovik, O.; Zeng, Z.C.; Yung, Y.L. Remote sensing impact of aerosol vertical distribution on aerosol optical depth retrieval from passive satellite sensors. *Remote Sens.* **2020**, *12*, 1524. [CrossRef]
11. Wood, R.; Bretherton, C.S. Boundary layer depth, entrainment, and decoupling in the cloud-capped subtropical and tropical marine boundary layer. *J. Clim.* **2024**, *17*, 3576–3588. [CrossRef]
12. Vakkari, V.; Baars, H.; Bohlmann, S.; Bühl, J.; Komppula, M.; Mamouri, R.E.; O'Connor, E.J. Aerosol particle depolarization ratio at 1565 nm measured with a Halo Doppler lidar. *Atmos. Chem. Phys.* **2021**, *21*, 5807–5820. [CrossRef]
13. Clark, N.E.; Pal, S.; Lee, T.R. Empirical evidence for the frontal modification of atmospheric boundary layer depth variability over land. *J. Appl. Meteor. Climatol.* **2022**, *61*, 1041–1063. [CrossRef]
14. Chu, Y.; Wang, Z.; Xue, L.; Deng, M.; Lin, G.; Xie, H.; Wang, Y. Characterizing warm atmospheric boundary layer over land by combining Raman and Doppler lidar measurements. *Opt. Express* **2022**, *30*, 11892–11911. [CrossRef] [PubMed]
15. Peerrone, M.R.; Lorusso, A.; Romano, S. Diurnal and nocturnal aerosol properties by AERONET sun-sky-lunar photometer measurements along four years. *Atmos. Res.* **2022**, *265*, 105889. [CrossRef]
16. Yang, H.; Qiu, D.; Fang, Z. LiDAR technology and experimental research for comprehensive measurement of atmospheric transmittance, turbulence, and wind. *J. Appl. Remote Sens.* **2024**, *18*, 12. [CrossRef]
17. Degnan, J.J. Evolution of Single Photon Lidar: From Satellite Laser Ranging to Airborne Experiments to ICESat-2. *Photonics* **2024**, *11*, 924. [CrossRef]
18. Chu, Y.; Lin, G.; Deng, M.; Guo, H.; Zhang, J.A. Characterizing Seasonal Variation of the Atmospheric Mixing Layer Height Using Machine Learning Approaches. *Remote Sens.* **2025**, *17*, 1399. [CrossRef]

19. Yang, Y.; Zhao, C.; Wang, Y. Multi-source Data Based Investigation of Aerosol-Cloud Interaction over the North China Plain and North of the Yangtze Plain. *J. Geophys. Res. Atmos.* **2021**, *19*, 126. [CrossRef]
20. Huang, Y.; Lu, X.; Fung, J.C.; Wong, D.C.; Li, Z.; Chen, Y.; Chen, W. Investigating southeast Asian biomass burning by the WRF-CMAQ two-way coupled model: Emission and direct aerosol radiative effects. *Atmos. Environ.* **2023**, *294*, 119521. [CrossRef] [PubMed]
21. Azmi, S.; Sharma, M.; Nagar, P.K. NMVOC emissions and their formation into secondary organic aerosols over India using WRF-Chem model. *Atmos. Environ.* **2022**, *287*, 119254. [CrossRef]
22. Bravo-Aranda, J.A.; Moreira, G.D.A.; Navas-Guzmán, F.; Granados-Muñoz, M.J.; Guerrero-Rascado, J.L.; Pozo-Vázquez, D.; Arbizu-Barrena, C.; Reyes, F.J.O.; Mallet, M.; Arboledas, L.A. A new methodology for PBL height estimations based on lidar depolarization measurements: Analysis and comparison against MWR and WRF model-based results. *Atmos. Chem. Phys.* **2017**, *17*, 6839–6851. [CrossRef]
23. Li, M.; Wu, Y.; Yuan, J.; Zhao, L.; Tang, D.; Dong, J.; Xia, H.; Dou, X. Stratospheric aerosol lidar with a 300m diameter superconducting nanowire single-photon detector at 1064 nm. *Opt. Express* **2023**, *31*, 2768–2779. [CrossRef] [PubMed]
24. Liu, X.; Zhang, L.; Zhai, X.; Li, L.; Zhou, Q.; Chen, X.; Li, X. Polarization Lidar: Principles and Applications. *Photonics* **2023**, *10*, 1118. [CrossRef]
25. Dai, G.; Wang, X.; Sun, K.; Wu, S.; Song, X.; Li, R.; Yin, J.; Wang, X. Calibration and retrieval of aerosol optical properties measured with Coherent Doppler Lidar. *J. Atmos. Ocean Technol.* **2021**, *38*, 1035–1045. [CrossRef]
26. Gong, W.; Zhang, J.; Mao, F.; Li, J. Measurements for profiles of aerosol extinction coeffcient, backscatter coeffcient, and lidar ratio over Wuhan in China with Raman/Mie lidar. *Chin. Opt. Lett.* **2010**, *8*, 533–536. [CrossRef]
27. Mallat, S.; Hwang, W. Singularity detection and processing with wavelets. *IEEE Trans. Inf. Theory* **1992**, *38*, 617–643. [CrossRef]
28. Morille, Y.; Haeffelin, M.; Drobinski, P.; Pelon, J. STRAT: An Automated Algorithm to Retrieve the Vertical Structure of the Atmosphere from Single-Channel Lidar Data. *J. Atmos. Ocean. Technol.* **2007**, *24*, 761–775. [CrossRef]
29. Zhou, T.; Huang, J.; Huang, Z.; Liu, J.; Wang, W.; Lin, L. The depolarization–attenuated backscatter relationship for dust plumes. *Opt. Express* **2013**, *21*, 15195–15204. [CrossRef] [PubMed]
30. Hersbach, H.; Bell, B.; Berrisford, P.; Biavati, G.; Horányi, A.; Muñoz Sabater, J.; Nicolas, J.; Peubey, C.; Radu, R.; Rozum, I.; et al. ERA5 Hourly Data on Single Levels from 1940 to Present. Copernicus Climate Change Service (C3S). 2023. Available online: https://cds.climate.copernicus.eu/datasets/reanalysis-era5-single-levels?tab=overview (accessed on 9 September 2023).
31. Shi, H.; Cao, X.; Li, Q.; Li, D.; Sun, J.; You, Z.; Sun, Q. Evaluating the Accuracy of ERA5 Wave Reanalysis in the Water Around China. *J. Ocean Univ. China* **2021**, *20*, 1–9. [CrossRef]
32. Sharmar, V.; Markina, M. Validation of global wind wave hindcasts using ERA5, MERRA2, ERA-Interim and CFSRv2 reanalyzes. *IOP Conf. Ser. Earth Environ. Sci.* **2020**, *606*, 012056. [CrossRef]
33. Fernald, F.G. Analysis of Atmospheric Lidar Observation: Some Comments. *Appl.Opt.* **1984**, *23*, 652–653. [CrossRef] [PubMed]
34. Vogelezang, D.H.P.; Holtslag, A.A.M. Evaluation and model impacts of alternative boundary-layer height formulations. *Bound.-Layer Meteorol.* **1996**, *81*, 245–269. [CrossRef]
35. Seidel, D.J.; Zhang, Y.; Beljaars, A.; Golaz, J.-C.; Jacobson, A.R.; Medeiros, B. Climatology of the planetary boundary layer over the continental United States and Europe. *J. Geophys. Res. Atmos.* **2012**, *117*, D17106. [CrossRef]
36. Guo, J.; Miao, Y.; Zhang, Y.; Liu, H.; Li, Z.; Zhang, W.; He, J.; Lou, M.; Yan, Y.; Bian, L.; et al. The climatology of planetary boundary layer height in China derived from radiosonde and reanalysis data. *Atmos. Chem. Phys.* **2016**, *16*, 13309–13319. [CrossRef]
37. Smalikho, I.N.; Banakh, V.A. Measurements of wind turbulence parameters by a 659 conically scanning coherent Doppler lidar in the atmospheric boundary layer. *Atmos. Meas. Tech.* **2017**, *660*, 4191–4208. [CrossRef]

photonics

MDPI

Article

Parts-per-Billion Detection of Hydrogen Sulfide via Cavity Ring-Down Spectroscopy

Wei Xu [1], Xuejun Wang [1], Lei Zhao [1], Jun Zou [1] and Bing Chen [2,*]

[1] Wuhan Second Ship Design and Research Institute, Wuhan 430205, China; 18007195900@163.com (W.X.); frison@163.com (X.W.); lemonredtea2024@163.com (L.Z.); zoujun1982@163.com (J.Z.)
[2] Anhui Institute of Optics and Fine Mechanics, Hefei Institutes of Physical Sciences, Chinese Academy of Sciences, Hefei 230031, China
* Correspondence: bchen@aiofm.ac.cn

Abstract: Rapid and precise detection of hydrogen sulfide (H_2S) at trace levels is critical for industrial safety and environmental air quality monitoring, yet existing methods often struggle with cost, speed, or sensitivity. A cost-effective cavity ring-down spectroscopy (CRDS) analyzer is presented, incorporating a novel digital locking circuit for sequential laser-cavity mode matching. This system demonstrates rapid and precise hydrogen sulfide (H_2S) detection capability at parts-per-billion (ppb) concentration levels. Compared to traditional wavelength meters, our system delivers a 140-fold improvement in frequency interval precision (0.07 MHz, 0.027% relative uncertainty). Allan variance analysis under vacuum conditions demonstrates a sensitivity limit of 3×10^{-12} cm^{-1} at a 60-s averaging time. Validated through calibrated gas dilution tests, the analyzer detects a 4 ppb H_2S absorption signal with a signal-to-noise ratio (SNR) > 6, establishing a 2 ppb detection limit (3σ criterion). This innovative approach meets stringent industrial and environmental requirements, offering a significant advancement in trace gas-sensing technology.

Keywords: cavity ringdown spectroscopy; hydrogen sulfide; trace gas detection

1. Introduction

Hydrogen sulfide (H_2S) is a critical trace gas with multifaceted applications spanning biomedical diagnostics, industrial safety, and energy infrastructure monitoring. Its role as a biomarker for pulmonary diseases (e.g., asthma and chronic obstructive pulmonary disease (COPD)) and as a decomposition byproduct of SF_6 insulation gas in high-voltage equipment necessitates ultrasensitive detection at parts-per-billion (ppb) levels [1]. In biomedical applications, H_2S functions as both a food spoilage biomarker via metabolic byproduct analysis and an endogenous gasotransmitter regulating cardiovascular physiology [2]. In respiratory medicine, sub-ppm H_2S concentrations in exhaled breath serve as biomarkers for early diagnosis of pulmonary diseases such as asthma and COPD [3].

The energy infrastructure sector demonstrates another crucial application of H_2S monitoring. As a decomposition byproduct of SF_6 insulation gas in high-voltage equipment, ppb-level H_2S detection provides critical predictive maintenance capabilities. Early identification of minute H_2S concentrations enables timely detection of insulation system degradation, thereby preventing catastrophic electrical failures in power transmission networks [4,5].

Industrial applications present particularly demanding challenges for H_2S monitoring technology. The thermal power generation industry exemplifies these requirements, where the widespread adoption of low-NOx combustion systems and circulating fluidized bed reactors has significantly elevated ambient H_2S concentrations. This heightened presence of reducing gases induces two major operational hazards: accelerated boiler coking/slagging processes and severe high-temperature sulfidation corrosion of water-cooled walls. The latter phenomenon poses particularly grave risks, as progressive wall thinning can culminate in catastrophic tube failures ("bursting"), potentially triggering major industrial accidents [6,7]. These challenges necessitate robust online monitoring systems capable of real-time, ultrasensitive H_2S detection under extreme industrial conditions.

These diverse demands have catalyzed the development of advanced sensing platforms capable of ppb-level sensitivity, high selectivity in multicomponent gas matrices, and resilience under harsh industrial conditions.

Recently, many techniques based on absorption spectroscopy of atoms and molecules in the gas phase have been developed in order to achieve a sensitive and accurate measurement of H_2S. Tunable diode laser absorption spectroscopy (TDLAS) techniques, including wavelength modulation and Herriot cell designs, have achieved detection limits of 140 ppb [8,9], while quantum cascade laser-based systems report 1.5 ppm sensitivity [10]. Fabry Perot (FP) interferometer is often used due to its excellent sensitivity and accuracy, and it also has been applied to detect low concentrations of H_2S [11,12]. In terms of quartz-enhanced photoacoustic spectroscopy (QEPAS), near-infrared QEPAS H_2S sensors that can achieve ppm level H_2S detection limits have been developed by some research teams [13–15]. Wu et al. achieved a detection limit of 142 ppb for H_2S in the 1582 nm band using an enhanced near-infrared QEPAS sensor [16]. A detection limit of 10 ppb for H_2S using double-resonant photoacoustic spectroscopy with line locking has been achieved by Zhang's group [17]. In terms of off-axis integrated cavity output spectroscopy, Tian et al. and Chen et al. have achieved a detection limit of 150 ppb for H_2S [18,19]. Despite advancements, existing CRDS systems face limitations in spectral acquisition speed and field-deployment robustness. For instance, Siciliani de Cumis et al. reported a 105 ppb detection limit with conventional CRDS, highlighting the need for enhanced sensitivity [20]. Jiang et al. achieved a detection limit of 4.44 ppb for H_2S in SF_6 gas using CRDS technology [21].

This study tackles these deficiencies by designing a miniaturized cavity ring-down spectroscopy (CRDS) system that integrates hybrid scanning algorithms and achieves sub-millikelvin (sub-mK) thermal stability, enabling parts-per-billion (ppb)-level detection of hydrogen sulfide (H_2S) in challenging and harsh environmental conditions. The proposed system overcomes the limitations in sensitivity, portability, and robustness found in conventional approaches, offering a compact and efficient solution for trace gas monitoring. The structure of this paper is organized as follows: Section 2 details the instrumentation and methodology, outlining the technical framework and experimental setup; Section 3 presents the results, providing a comprehensive analysis of the system's performance; the final section offers a conclusion and discussion, synthesizing key findings, implications, and potential avenues for future research.

2. Experimental Setup and Method

2.1. Experimental Setup

As illustrated in the schematic diagram of the cavity ring-down spectroscopy (CRDS) system, the DFB laser (GyScienTech, Wuhan City, China, GST1574DFBBF1413-PM) emits a wavelength-tunable beam modulated by an acousto-optic modulator (AOM). The modulated light is coupled into a quartz ring-down cavity (effective optical path: 660 mm; baseline ring-down time: 191.5 μs) via a single-mode polarization maintaining fiber.

High-precision hydrogen sulfide (H_2S) detection is fundamentally constrained by three critical parameters: thermodynamic stability, pressure regulation, and spectral interference mitigation. To address these challenges, our apparatus employs a dual-stage thermal control system that stabilizes intracavity gas temperature at 45.00 °C ± 0.005 °C, achieving sub-millikelvin thermal uniformity. Under ambient pressure conditions, trace-level H_2S concentrations exhibit significant spectral overlap interference from dominant CO_2 and H_2O absorption lines, though such cross-talk effects demonstrate progressive attenuation with reduced pressure.

During dynamic pressure optimization, the minimum achievable cavity pressure is constrained to >20 kPa due to cumulative instrumental tolerances in the vacuum pump assembly, mass flow controller (200 sccm ± 0.01 sccm), and proportional control valve. To ensure robust pressure regulation with adequate system margin, the operational pressure is actively maintained at 25.000 kPa ± 0.002 kPa through closed-loop control. This optimized pressure regime enables unambiguous molecular discrimination of H_2S signatures.

Wavelength selection is controlled by a microcontroller unit (STM32H743) interfaced with a 16-bit digital-to-analog converter (AD5683, Analog Devices), which modulates the laser's temperature and injection current. The amplified photodetector output is digitized using an 18-bit analog-to-digital converter (Analog Devices, AD4003) at a sampling rate of 2 MSPS, ensuring high-fidelity ring-down waveform capture. Upon threshold detection, the MCU generates a TTL signal to deactivate the AOM, ensuring optimal ring-down waveform acquisition. Following the acquisition of complete ring-down waveforms, the MCU transmits spectral data to a host computer for real-time concentration inversion and graphical display via custom-developed algorithms as shown in Figure 1.

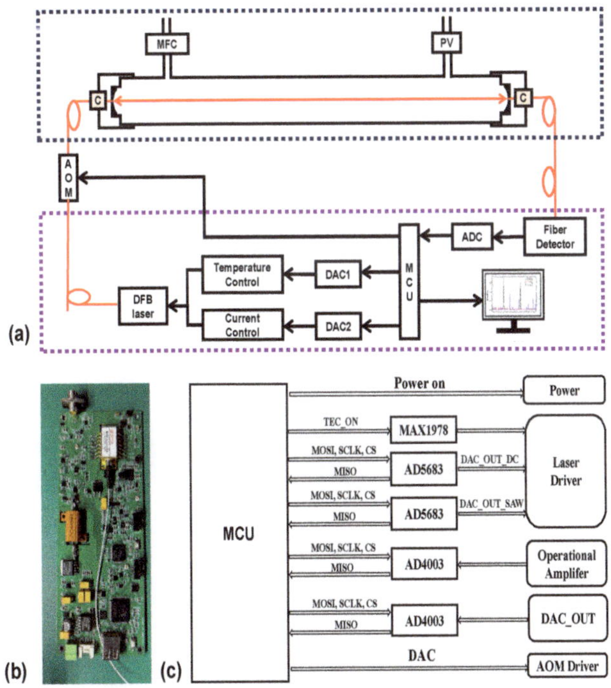

Figure 1. Schematic of the cavity ring-down spectroscopy analyzer (CRDS) for H_2S detection. (**a**) Schematic diagram of the experimental apparatus. MFC: mass flow controllers; PV: proportional valve; C: self-made collimator; AOM: acousto-optic modulator; DAC: digital-to-analog converter; ADC: analog-to-digital converter; MCU: microcontroller unit. (**b**) Photograph of the control circuit. (**c**) Schematic diagram of the control circuit.

2.2. Spectral Scan Method

The longitudinal modes of the ring-down cavity serve as precise frequency markers, with their frequencies expressed as follows:

$$\nu_N = \nu_0 + N \times FSR \quad (1)$$

where N is the mode order and FSR is the free spectral range. Enhanced FSR stability is achieved through two key measures: (1) employing low-expansion-coefficient materials (high-purity quartz in this work) and (2) implementing precision temperature control. Here, ν_0 corresponds to the fundamental mode frequency (N = 0) and ν_N specifies the Nth-order mode frequency.

While the sequential matching of cavity modes could theoretically enable complete spectral acquisition, practical limitations arise from laser current–wavelength nonlinearities and frequency instability. Our previous digital locking schemes [22] improved mode matching efficiency but exhibited prolonged relocking durations (>10 s) following unexpected disturbances. To address this limitation, we developed a hybrid scanning strategy combining wide-range searching (3/2 FSR span) with narrow-band locking (1/5 FSR modulation), achieving spectral acquisition within 3–5 s. The operational workflow (see schematic diagram as shown in Figure 2) comprises the following:

1. Coarse Search Phase: Detect mode-matching conditions through 3/2 FSR scanning.
2. Precision Locking Phase: Apply 1/5 FSR modulation for stabilized measurements.
3. Fault Recovery: Reinitiate searching mode if no ring-down signal is detected within 0.1 s.

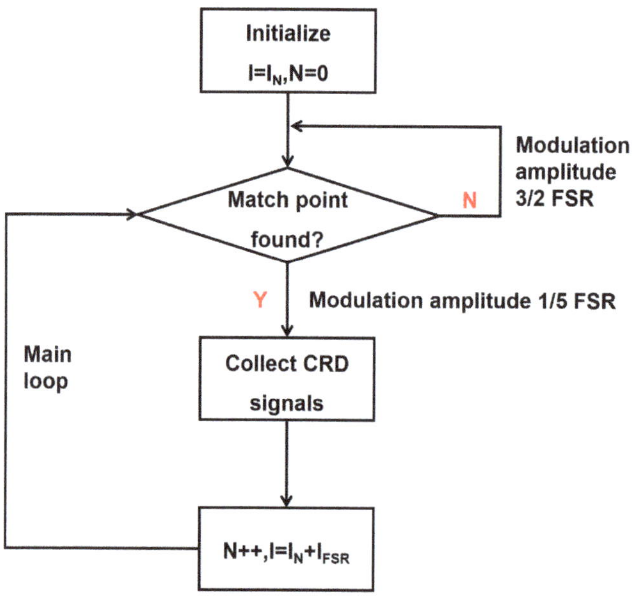

Figure 2. Control diagram for a strict mode-by-mode scan.

This adaptive approach maintains measurement continuity while preventing prolonged system downtime from transient disturbances.

3. Results

3.1. Self-Calibration of the Spectrum

To assess the accuracy of the longitudinal mode frequency comb, we conducted continuous spectral measurements of H_2S absorption features near 1574.554 nm over a 5 h period. Through multi-peak fitting analysis, we determined the relative positions of three characteristic absorption peaks. These measured peak positions were then correlated with reference wavelengths from the HITRAN database using linear regression analysis. The slope of the fitted curve yielded the free spectral range (FSR) of 257.26 $\pm$ 0.07 MHz.

The measurement precision was further evaluated through statistical analysis of 76 discrete spectral data points. The maximum potential relative error for any individual point is theoretically 1/76 (1.31%). However, as shown in Figure 3b, the observed maximum relative deviation was only 0.18%, demonstrating two critical findings:

- The scanning methodology achieved complete spectral coverage without missing data points.
- The measurement precision significantly exceeded the theoretical minimum requirement.

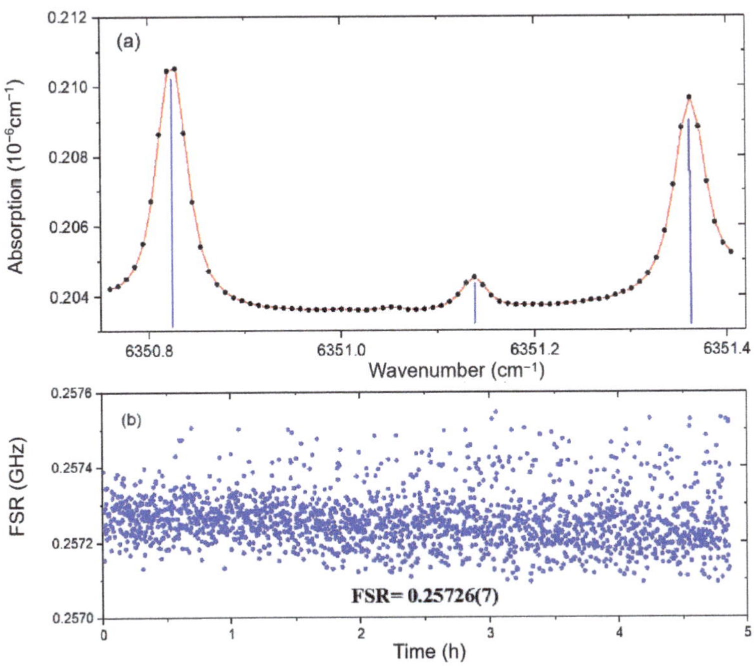

Figure 3. (**a**) Absorption spectrum of H_2S near 6351.0 cm^{-1} at 250 kPa. (**b**) Statistical distribution of FSR measurements over 5 h continuous measurement.

This exceptional performance validates the robustness of our hybrid scanning approach and the stability of the frequency comb reference.

3.2. Linearity, Sensitivity, Detection Limit, and Response Time

Instrument calibration was performed using four certified reference gas cylinders (Nanjing Special Gases Co., Ltd.) with certified H_2S concentrations of N_2 blank, 5 ppm, 18.1 ppm, and 50.2 ppm (1% uncertainty). Hydrogen sulfide concentrations were subsequently determined using cavity ring-down spectroscopy (CRDS). As shown in Figure 4, the CRDS measurements exhibited excellent agreement with the certified gravimetric mixture (GM) values, demonstrating an Adj. R-Square of 0.99959. The observed congruence between CRDS and GM measurements confirms that the cavity ring-down spectrometer has achieved satisfactory quantitative accuracy for precise concentration determinations.

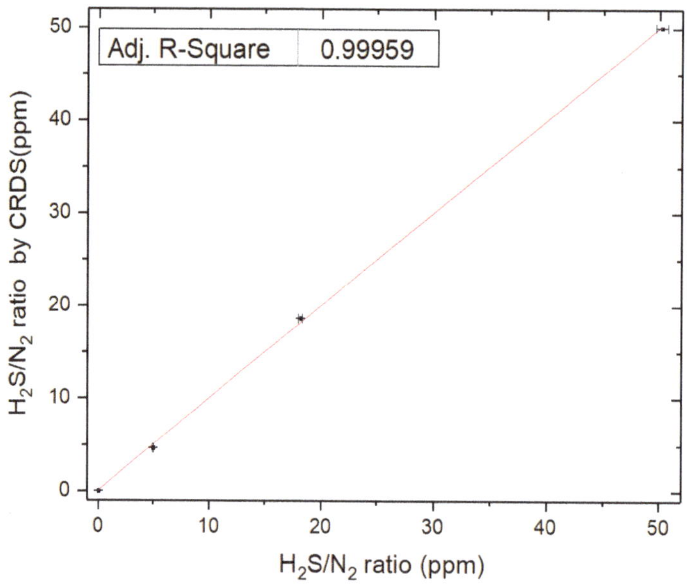

Figure 4. H_2S concentration in N_2 determined by CRDS.

Sensitivity, defined as the minimum resolvable concentration variation of the instrument, is governed by three critical parameters: (1) the signal-to-noise ratio (SNR) of the detection system, (2) the quality factor (Q-value) of the optical cavity, and (3) the stability of the laser source. Allan variance analysis, which evaluates the evolution of signal noise as a function of integration time, serves as a robust tool to determine the optimal integration time and derive the minimum detection limit (MDL)—the lowest H_2S concentration that can be reliably identified at a specified confidence level. Compared to conventional noise analysis methods, Allan variance provides distinct advantages:

- Noise Discrimination: Differentiates contributions from white noise, flicker noise, and other noise types.
- Detection Limit Optimization: Enables significant reduction of MDL through systematic integration time tuning.
- Long-Term Stability Quantification: Offers a quantitative framework for assessing system performance in trace gas detection applications.

To evaluate the detection capability of our system, we performed continuous measurements of cavity ring-down time variations at a fixed wavelength within the H_2S absorption band (1574.554 nm) under vacuum conditions (Figure 5a). Allan variance analysis revealed a minimum detectable absorption coefficient of 3×10^{-12} cm^{-1} at 60-s integration (Figure 5b), corresponding to 0.2 ppb H_2S concentration fluctuations—surpassing the sensitivity of prior CRDS systems. This performance highlights the exceptional stability and precision of the developed CRDS system.

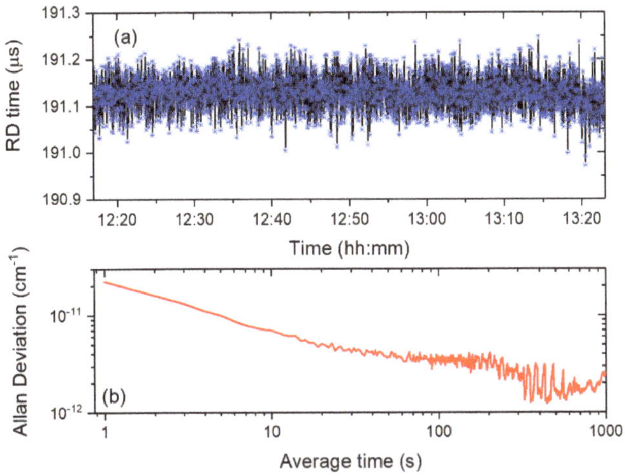

Figure 5. Cavity loss ($1/c\tau$) of an empty RD cavity (upper panel (**a**)) and the Allan deviation (lower panel (**b**)).

To establish a controlled low-concentration H_2S environment, we implemented a dynamic dilution protocol using a certified 4 ppm H_2S/air mixture. The experimental sequence comprised the following:

- Initial Conditioning: Purging the cavity with 4 ppm H_2S reference gas.
- Dynamic Dilution: Continuous air injection at constant flow (200 sccm) and pressure (25,000 Pa).
- Signal Processing: Application of moving average filtering (3 s per spectrum, 100-point window) for enhanced measurement precision.

As illustrated in Figure 6, a dynamic dilution system employing purified air was utilized to systematically attenuate hydrogen sulfide (H_2S) standard gas at a constant flow rate of 200 standard cubic centimeters per minute (sccm). The experimentally monitored temporal concentration profile demonstrated sequential attenuation characteristics: the H_2S concentration decreased from an initial 4 ppm to 88 ppb after 10 min of continuous dilution, followed by subsequent reductions to 27 ppb at 20 min, 10 ppb at 30 min, and ultimately reaching 4 ppb after 45 min of operation.

The absorption spectra exhibited a signal-to-noise ratio (SNR) exceeding 6 across the tested concentration range (4–80 ppb), validating the system's robustness for trace-level quantification under dynamic dilution conditions. Applying the 3σ detection criterion, the system achieves a practical detection limit of 2 ppb in ambient air, validating its capability for air trace gas monitoring applications.

Instrument response characterization was conducted through a gas-switching methodology employing H_2S standard gas and high-purity nitrogen (N_2). This protocol was implemented to determine the temporal response characteristics and establish the detection limits under standardized gas conditions.

The temporal resolution was quantitatively evaluated using two critical parameters: T_{90} and T_{10} response times. The operational definitions were as follows:

- T_{90}: Duration required for sensor output to reach 90% of nominal value during the transition from high-purity N_2 to calibrated H_2S concentration.
- T_{10}: Time elapsed for signal decay to 10% of initial value when switching from H_2S standard gas back to high-purity N_2.

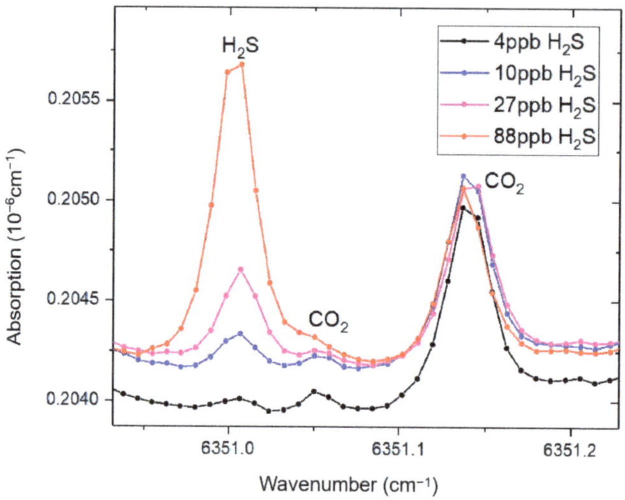

Figure 6. Temporal profile of H_2S concentration during dynamic dilution tests. The concentration decreased from 80 ppb to 4 ppb over 45 min, with absorption spectra recorded at each dilution step.

As shown in Figure 7, the system exhibited a T_{90} response time of 106 s and a T_{10} recovery time of 85 s, reflecting efficient gas-switching dynamics. The main reason for the difference in response time is that we adopt the sliding average method, which improves the detection ability of the system but reduces the response of the system, especially the rise time response T90. Subsequent statistical analysis of the baseline signal following prolonged N_2 purging revealed a standard deviation of 0.6 ppb for H_2S concentrations, corresponding to a measurement precision of 0.6 ppb. Using the conventional three-sigma criterion, the instrument demonstrated a detection limit of 1.8 ppb.

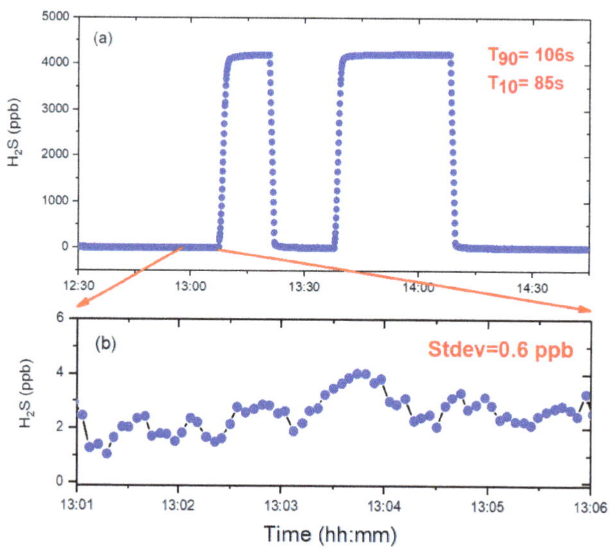

Figure 7. (**a**) Temporal response of the CRDS system during gas-switching cycles between 4 ppm H_2S and N_2, showing T_{90} = 106 s and T_{10} = 85 s, (**b**) Close-up visualization of the region of interest (ROI) marked in (**a**).

4. Conclusions and Discussion

This study introduces a low-cost, high-precision cavity ring-down spectroscopy (CRDS) analyzer tailored for trace hydrogen sulfide (H_2S) detection, addressing the pressing need for sensitive, affordable, and field-deployable gas-sensing technologies. The system integrates a novel digital locking circuit within a compact, embedded platform that seamlessly combines laser frequency stabilization, ring-down signal detection, amplification, data acquisition, exponential fitting, and real-time data transmission. Spectral analysis and visualization are efficiently handled by a dedicated microcomputer, streamlining the operational workflow. By adopting an all-polarization-maintaining-fiber configuration, the design eliminates free-space optical components, enhancing mechanical stability and reducing maintenance demands. The resulting analyzer, with dimensions of 48 cm × 22 cm × 95 cm and a power consumption of 70 W, offers a practical solution for both laboratory and field applications.

The developed H_2S sensing system demonstrates unprecedented sensitivity, attaining a record-low detection limit of 1.8 ppb. A systematic comparison against existing detection methods reveals significant performance advantages; the details are provided in Table 1.

Table 1. Comparison of the different H_2S detection techniques.

Techniques	Spectral Regions	Detection Limit
WMS-TDLAS + 15 m Herriott multipass cell [8]	Near-IR	140 ppb
QCL-TDLAS + 100 m Herriott multipass cell [10]	Mid-IR	1500 ppb
QEPAS + Fiber amplifier [14]	Near-IR	142 ppb
QEPAS + Line locking [17]	Near-IR	10 ppb
CRDS [20]	Near-IR	105 ppb
CRDS (this work)	Near-IR	2 ppb

This work represents a 52.5-fold improvement over the 105 ppb limit of existing field-deployable CRDS systems [20]. This leap in performance stems from two key innovations: a hybrid scanning algorithm that optimizes laser-cavity mode alignment and a thermally stabilized cavity (±0.005 °C) that minimizes frequency drift. The detection scheme exploits the stable longitudinal modes of the ring-down cavity as an intrinsic frequency comb, paired with a novel digital locking algorithm that sequentially aligns the laser frequency to individual cavity modes with unprecedented precision. A 5 h characterization of the free spectral range (FSR) demonstrated a frequency precision of 0.07 MHz (relative precision: 0.02%), far surpassing the ~10 MHz resolution of conventional wavelength meters. This advancement not only enhances the accuracy of H_2S absorption line measurements but also sets a new benchmark for frequency stabilization in CRDS systems.

Sensitivity assessments further underscore the system's capabilities. Allan variance analysis of cavity ring-down times under vacuum conditions revealed a minimum detectable absorption coefficient of 3×10^{-12} cm^{-1} at a 60 s integration time, reflecting the analyzer's ability to detect ultra-low gas concentrations with high stability. Field deployment validated its real-world efficacy, achieving a detection limit below 2 ppb for atmospheric H_2S with a T_{90} response time of 106 s and a signal-to-noise ratio (SNR) exceeding 6 for a 4 ppb signal. These metrics fulfill stringent requirements for industrial safety protocols and environmental air quality monitoring, where rapid and reliable detection of trace H_2S is paramount.

The modular design of this CRDS analyzer enhances its versatility, positioning it as a scalable platform for monitoring multiple greenhouse gases and polar molecules beyond H_2S. The incorporation of adsorption-resistant modifications—such as specialized coatings or materials—mitigates the signal degradation caused by gas adhesion to cavity surfaces, a common challenge in trace gas sensing. Preliminary tests suggest that, with targeted surface modifications, the system could be adapted for real-time detection of species like ammonia (NH_3) and so on [23], expanding its utility in climate research, agricultural monitoring, and industrial process control. The compact footprint, low power consumption, and cost-effective architecture further distinguish this platform from bulkier, more resource-intensive alternatives, making it an attractive option for widespread deployment.

Despite these achievements, opportunities for refinement remain. The T90 response time of 106 s, while adequate for many applications, could be reduced through faster signal processing or enhanced cavity flushing mechanisms to meet the demands of dynamic, high-throughput environments. Additionally, while the current design excels in controlled settings, its performance under variable humidity, temperature, or pressure conditions warrants further investigation to ensure robustness across diverse field scenarios. Future work could also explore integrating multi-wavelength capabilities to simultaneously detect multiple gas species, amplifying the system's analytical power.

In summary, this work demonstrates a transformative approach to CRDS-based trace gas detection, blending affordability, precision, and adaptability into a single platform. By leveraging innovative digital locking techniques and a streamlined optical design, the analyzer not only outperforms existing systems in H_2S detection but also lays the groundwork for broader applications in environmental and industrial monitoring. These advancements herald a step forward in the evolution of portable, high-sensitivity spectroscopy tools, with significant potential to address emerging challenges in gas-sensing technology.

Author Contributions: Conceptualization, W.X. and X.W.; methodology, W.X. and B.C.; software, J.Z.; validation, L.Z. and X.W.; formal analysis, W.X.; investigation, J.Z.; writing—original draft preparation, W.X.; writing—review and editing, B.C. All authors have read and agreed to the published version of the manuscript.

Funding: This research was funded by the National Natural Science Foundation of China (41901081).

Institutional Review Board Statement: Not applicable.

Informed Consent Statement: Not applicable.

Data Availability Statement: The data that support the plots within this paper are available from the corresponding author on request basis.

Conflicts of Interest: The authors declare no conflicts of interest.

References

1. Wang, R. Physiological implications of hydrogen sulfide: A whiff exploration that blossomed. *Physiol. Rev.* **2012**, *92*, 791–896. [CrossRef] [PubMed]
2. Kimura, H. Hydrogen sulfide as a neuromodulator. *Mol. Neurobiol.* **2002**, *26*, 13–19. [CrossRef] [PubMed]
3. Suzuki, Y.; Saito, J.; Munakata, M.; Shibata, Y. Hydrogen sulfide as a novel biomarker of asthma and chronic obstructive pulmonary disease. *Allergol. Int.* **2021**, *70*, 181–189. [CrossRef]
4. Lv, H.; Zhang, X.; Jiang, A.; Qian, W.; Zhang, C.; Zhang, X. Detection of SF_6 decomposition components H_2S and CO_2 based on WDM and CRDS. *IEEE Trans. Dielectr. Electr. Insul.* **2024**, *31*, 2922–2929. [CrossRef]
5. Gao, R.; Zhu, M.; Zhang, C. The uncertainty evaluation of determining the trace H_2S in SF_6/N_2 insulating gas mixture by cavity ring-down spectroscopy. *J. Phys. Conf. Ser.* **2024**, *2774*, 012030. [CrossRef]
6. Sun, X.; Ning, Y.; Yang, J.; Zhao, Y.; Yang, Z.; Zhou, X. Study on high temperature corrosion mechanism of water wall tubes of 350 MW supercritical unit. *Eng. Fail. Anal.* **2021**, *121*, 105131. [CrossRef]
7. Wang, Y.; Li, L.; Wang, M.; Tan, H.; Zhang, S. Effect of ZnS/PbS deposits on high temperature corrosion of watewall tubes in reducing atmosphere. *Fuel Process. Technol.* **2021**, *216*, 106793. [CrossRef]
8. Guo, Y.; Qiu, X.; Li, N.; Feng, S.; Cheng, T.; Liu, Q.; He, Q.; Kan, R.; Yang, H.; Li, C. A portable laser-based sensor for detecting H_2S in domestic natural gas. *Infrared Phys. Techn.* **2020**, *105*, 103153. [CrossRef]

9. Xia, H.; Dong, F.; Wu, B.; Zhang, Z.; Pang, T.; Sun, P.; Cui, X.; Han, L.; Wang, Y. Sensitive absorption measurements of hydrogen sulfide at 1.578 μm using wavelength modulation spectroscopy. *Chinese Phys. B* **2015**, *24*, 034204. [CrossRef]
10. Moser, H.; Genner, A.; Ofner, J.; Schwarzer, C.; Strasser, G.; Lendl, B. Application of a ring cavity surface emitting quantum cascade laser (RCSE-QCL) on the measurement of H_2S in a CH_4 matrix for process analytics. *Opt. Express* **2016**, *24*, 6572–6585. [CrossRef]
11. Chu, Y.; Alshammari, M.; Wang, X.; Han, M. Angle-tunable method for optimizing rear reflectance in Fabry–Perot interferometers and its application in fiber-optic ultrasound sensing. *Photonics* **2024**, *11*, 1100. [CrossRef]
12. Kitture, R.; Pawar, D.; Rao, C.N.; Choubey, R.K.; Kale, S.N. Nanocomposite modified optical fiber: A room temperature, selective H_2S gas sensor: Studies using ZnO-PMMA. *J. Alloys Compd.* **2017**, *695*, 2091–2096. [CrossRef]
13. Viciani, S.; Siciliani de Cumis, M.; Borri, S.; Patimisco, P.; Sampaolo, A.; Scamarcio, G.; De Natale, P.; D'Amato, F.; Spagnolo, V. A quartz-enhanced photoacoustic sensor for H_2S trace-gas detection at 2.6 μm. *Appl. Phys. B* **2015**, *119*, 21–27. [CrossRef]
14. Wu, H.; Sampaolo, A.; Dong, L.; Patimisco, P.; Liu, X.; Zheng, H.; Yin, X.; Ma, W.; Zhang, L.; Yin, W.; et al. Quartz enhanced photoacoustic H_2S gas sensor based on a fiber-amplifier source and a custom tuning fork with large prong spacing. *Appl. Phys. Lett.* **2015**, *107*, 111104. [CrossRef]
15. Olivieri, M.; Menduni, G.; Giglio, M.; Sampaolo, A.; Patimisco, P.; Wu, H.; Dong, L.; Spagnolo, V. Characterization of H_2S QEPAS detection in methane-based gas leaks dispersed into environment. *Photoacoustics* **2023**, *29*, 100438. [CrossRef]
16. Wu, H.; Dong, L.; Zheng, H.; Liu, X.; Yin, X.; Ma, W.; Zhang, L.; Yin, W.; Jia, S.; Tittel, F.K. Enhanced near-infrared QEPAS sensor for sub-ppm level H_2S detection by means of a fiber amplified 1582 nm DFB laser. *Sensor. Actuat. B: Chem.* **2015**, *221*, 666–672. [CrossRef]
17. Zhang, H.; Wang, Z.; Wang, Q.; Borri, S.; Galli, I.; Sampaolo, A.; Patimisco, P.; Spagnolo, V.L.; Natale, P.D.; Ren, W. Parts-per-billion-level detection of hydrogen sulfide based on doubly resonant photoacoustic spectroscopy with line-locking. *Photoacoustics* **2023**, *29*, 100436. [CrossRef]
18. Tian, X.; Cheng, G.; Cao, Y.; Chen, J.; Liu, K.; Gao, X. Simultaneous detection of hydrogen sulfide and carbon dioxide based on off-axis integrated cavity output spectroscopy using a near-infrared distributed feedback diode laser. *Microw. Opt. Techn. Let.* **2021**, *63*, 2074–2078. [CrossRef]
19. Chen, W.; Kosterev, A.A.; Tittel, F.K.; Gao, X.; Zhao, W. H_2S trace concentration measurements using off-axis integrated cavity output spectroscopy in the near-infrared. *Appl. Phys. B* **2008**, *90*, 311–315. [CrossRef]
20. Siciliani de Cumis, M.; Viciani, S.; Galli, I.; Mazzotti, D.; Sorci, F.; Severi, M. Note: An analyzer for field detection of H_2S by using cavity ring-down at 1.57 μm. *Rev. Sci. Instrum.* **2015**, *86*, 056108. [CrossRef]
21. Jiang, A.; Zhang, C.; Zhu, M.; Zhang, X. Detection of characteristic component H_2S in SF_6 and SF_6/N_2 mixture using cavity ring-down spectroscopy. *High Volt.* **2025**, *10*, 88–94. [CrossRef]
22. Chen, B.; Sun, Y.R.; Zhou, Z.; Chen, J.; Liu, A.; Hu, S. Ultrasensitive, self-calibrated cavity ring-down spectrometer for quantitative trace gas analysis. *Appl. Optics* **2014**, *53*, 7716–7723. [CrossRef] [PubMed]
23. Gadedjisso-Tossou, K.S.; Stoychev, L.I.; Mohou, M.A.; Cabrera, H.; Niemela, J.; Danailov, M.B.; Vacchi, A. Cavity ring-down spectroscopy for molecular trace gas detection using a pulsed DFB QCL emitting at 6.8 μm. *Photonics* **2020**, *7*, 74. [CrossRef]

photonics

MDPI

Article

Enhancement of Methane Detection in Tunable Diode Laser Absorption Spectroscopy Using Savitzky–Golay Filtering

Shichao Chen [1,2,*], Xing Tian [1,2,3,*], Tong Mu [1,2], Jun Yuan [1,2], Xile Cao [1,2] and Gang Cheng [3]

1 The First Affiliated Hospital of Anhui University of Science and Technology (Huainan First People's Hospital), Anhui University of Science and Technology, Huainan 232001, China; 15178219810@163.com (T.M.); yuanjun1186196986@163.com (J.Y.); caoxile2024624@163.com (X.C.)
2 School of Artificial Intelligence, Anhui University of Science and Technology, Huainan 232001, China
3 State Key Laboratory of Mining Response and Disaster Prevention and Control in Deep Coal Mines, Anhui University of Science and Technology, Huainan 232001, China; chgmech@mail.ustc.edu.cn
* Correspondence: csc18856683605@163.com (S.C.); 2019018@aust.edu.cn (X.T.)

Abstract: In order to enhance gas absorption efficiency and improve the detection sensitivity of methane, a gas absorption cell with an effective optical path length of 29.37 m was developed, employing tunable diode laser absorption spectroscopy (TDLAS) and a distributed feedback (DFB) laser with a center wavelength of 1.654 μm as the light source. However, despite these advancements, the detection accuracy was still limited by potential signal interference and noise. To address these challenges, the Savitzky–Golay (S-G) filtering technique was implemented to optimize the TDLAS detection signal. Experimental results indicated a significant enhancement in detection performance. For a methane concentration of 92 ppm, the application of the S-G filter improved the signal-to-noise ratio by a factor of 1.84, resulting in a final device detection accuracy of 0.53 ppm. This improvement demonstrates the effectiveness of the S-G filter in enhancing detection sensitivity, supporting high-precision methane monitoring for atmospheric analysis and various industrial applications.

Keywords: TDLAS; Savitzky–Golay; signal-to-noise ratio; detection limit

1. Introduction

With the growing emphasis on environmental protection and renewable energy worldwide, methane, as a potent greenhouse gas, has garnered significant attention due to its role in global warming and climate change. Methane emissions, primarily originating from energy conversion processes, industrial production, and agricultural activities, have a substantial impact on climate dynamics, thereby driving an increasing need for effective and real-time monitoring of methane concentrations [1–3].Traditional methane detection methods, such as gas chromatography and catalytic combustion, often face challenges due to constraints imposed by environmental factors, including variations in temperature and humidity, as well as limitations in equipment sensitivity and response time. These limitations can significantly hinder the ability to achieve high-precision and high-sensitivity methane detection, especially in dynamic or harsh environmental conditions. Moreover, traditional methods typically require complex sample pretreatment, which limits their applicability for real-time and in situ monitoring purposes [4].

In light of these challenges, the development and refinement of novel methane detection technologies have become crucial. Tunable diode laser absorption spectroscopy (TDLAS), characterized by its high sensitivity, selectivity, and fast response, has emerged as a leading technique for methane detection in complex environments. These characteristics render TDLAS a powerful tool for gas analysis, exploiting the unique infrared absorption spectra of target molecules for precise and reliable measurements. Over the past few years, substantial research has showcased the effectiveness of TDLAS for methane detection, underscoring its potential in achieving high-precision environmental monitoring and advancing related fields [5–7]. In the TDLAS, the optical multi-pass absorption cell is one of the key equipment. The design principle of the multi-pass absorption cell is based on limiting the incident light to reflect back and forth between two or more highly reflective mirrors, thereby increasing the effective optical path of the light–substance interaction to obtain the absorption spectral signal with a high signal-to-noise ratio (SNR), thus realizing high sensitivity measurement of trace gases at low concentrations. In a certain volume and physical size of the optical multi-pass absorption cell, the number of reflections is increased as much as possible to improve the effective optical path, but because they are easily affected by light overlap resulting in light path interference, the reflections will ultimately affect the detection sensitivity and accuracy of the measurement system. Additionally, noise generated by sensors, detectors, and other electronic devices can increase the uncertainty in the signal processing process. Even with high-precision instruments, the detection of low-concentration methane may still be limited by noise. These limitations seriously hinder the ability to achieve high precision and sensitivity in methane detection, especially under dynamic or harsh environmental conditions [8–11].

Addressing these challenges requires effective signal processing and optimization strategies, which have become the focus of research in gas sensing technology. One such promising technique is the Savitzky–Golay filter, which is a powerful signal smoothing algorithm widely used in data analysis [12,13]. Compared with conventional filters such as moving average filters and low-pass filters, Savitzky–Golay filters can well retain the basic features of the original signal, such as peak shape and peak width, and also effectively reduce the noise and improve the signal-to-noise ratio (SNR). It is therefore particularly suitable for quantitative spectroscopic analysis where maintaining signal integrity is essential for accurate concentration calculations. The Savitzky–Golay filter technique has been successful in a number of fields including analytical chemistry, biomedical signal analysis, and environmental monitoring, providing a valuable method for the detection of methane using TDLAS. The integration of this filtering technique into the methane TDLAS detection framework has been shown to be effective in improving the measurement accuracy and reliability [14–16]. For instance, Ji et al. [17] combined wavelet denoising (WD) and Savitzky–Golay (SG) to improve the performance of a methane sensor to process the noisy signal, and the SNR of the absorbance signal was improved by a factor of 4.95 with the wavelet denoising and Savitzky–Golay filtering scheme. The 1 sigma limit of detection of the sensor is reduced from 2929 ppm to 397 ppm, according to the Allan deviation analysis. By mitigating environmental and instrumental noise, the method improves the SNR, allowing for a lower detection limit and increased sensitivity. Such advancements are vital for real-time environmental monitoring and industrial safety, where high precision and reliability are essential. Thus, incorporating advanced signal processing techniques, such as the Savitzky–Golay filter, is a significant step towards optimizing TDLAS-based methane detection, ultimately leading to more accurate and effective monitoring of methane emissions.

This study presents the development of a methane detection system based on tunable diode laser absorption spectroscopy, incorporating a novel, compact, dense-pattern multi-pass gas cell with an effective optical path length of 29.37 m. A near-infrared distributed feedback (DFB) diode laser operating at a wavelength of 1.654 μm was utilized as the light source. To further enhance the detection performance, the Savitzky–Golay (S-G) filtering technique was employed, leading to an improvement in the signal-to-noise ratio (SNR) by a factor of 1.84. Consequently, the system has a detection accuracy of 0.53 ppm when measuring a methane concentration of 92 ppm, indicating its potential for high-precision methane monitoring in both environmental and industrial applications.

2. Experimental Details

A schematic diagram of the experimental setup is shown in Figure 1. The light source for the system is a distributed feedback (DFB) diode laser operating at a wavelength of 1.654 μm and at room temperature. To ensure stable and precise operation, the laser current and temperature were controlled using a commercial diode laser controller (Model LDC 501, Stanford Research Systems, Sunnyvale, CA, USA), which maintained a constant operating temperature of 22 °C and a driving current of 85 mA. To achieve wavelength scanning, an external voltage ramp from a function generator (DG4162, RIGOL, Beijing, China) was applied to the laser diode current. This ramp modulated the laser wavelength back and forth across the absorption line with a peak amplitude of 1 V and a scanning frequency of 100 Hz, allowing precise control of the laser emission over the desired spectral range. The laser beam was collimated with a fiber-coupled collimator and subsequently injected into the compact dense-pattern multi-pass gas cell. Inside the multi-pass cell, the laser beam was reflected 243 times between two concave mirrors, resulting in an effective optical path length of 29.37 m, which significantly enhances the sensitivity of the system by increasing the interaction time between the laser and the target gas [18]. After passing through the gas cell, the output beam was focused onto a photodetector (PDA20CS-EC, Thorlabs, Newton, NJ, USA) using a lens with a focal length of 50 mm. The photodetector converted the optical signal into an electrical signal, which was then digitized by a data acquisition (DAQ) card (NI USB-6210, National Instruments, Austin, TX, USA). The acquired data were displayed and analyzed in real-time on a laptop using a LabVIEW NXG 5.0 interface, providing immediate feedback for monitoring and further analysis.

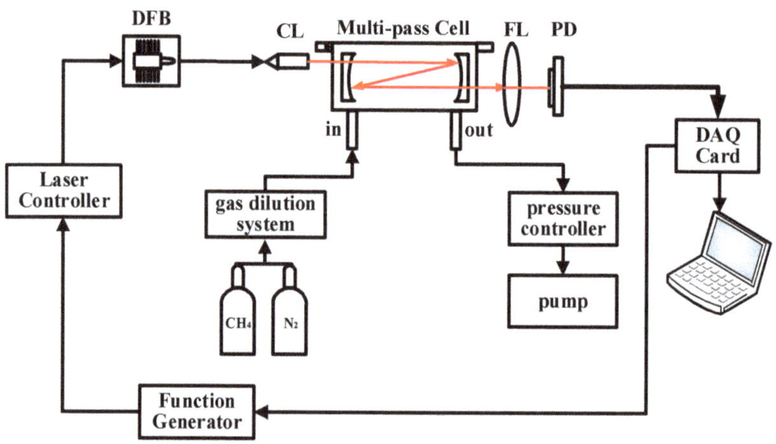

Figure 1. Schematic diagram of experimental device of measurement system. CL: collimator; FL: focusing lens; PD: photodetector.

3. Results and Discussion

3.1. Selection of CH_4 Absorption Lines

According to the HITRAN 2016 database, the best CH_4 absorption lines within the spectral tuning range of the used 1.654 μm diode laser are the R3 triplet of the $2v_3$ band near 6046.96 cm^{-1}. These lines exhibit the highest absorption intensities and are free from significant interference by other molecules, such as H_2O and CO_2. In the experimental setup, the three individual spectral lines of the R3 triplet are approximated as a single spectral line for processing purposes. The line strength used for analysis is taken as the average of the absorption strengths of these three lines [19,20]. Figure 2 shows simulated absorption spectra of 492 ppm CH_4 near the 6046.96 cm^{-1} at a pressure of 0.5 atm. The laser center current was set at 85 mA with a temperature set at 22 °C for targeting the selected CH_4 absorption lines. The figure also illustrates the relationship between the applied current and the corresponding wavenumber, which is crucial for calibrating the system and ensuring accurate methane concentration measurements.

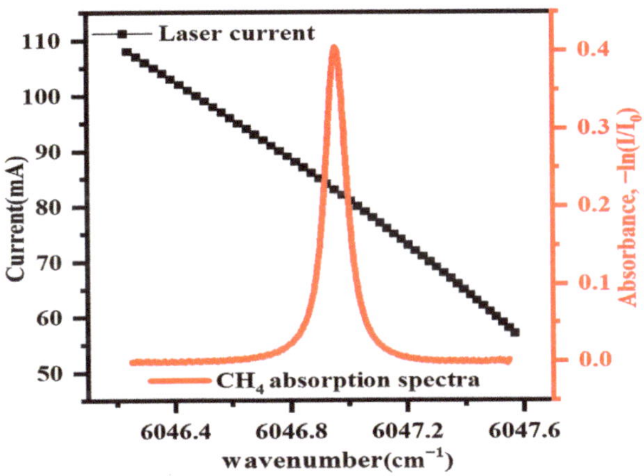

Figure 2. The simulated CH_4 absorption coefficient spectrum and the relationship between the laser injection current and the wavenumber.

3.2. The Measured Signal of CH_4

In the experiment, five groups of CH_4 gases with different concentrations of 92 ppm, 192 ppm, 292 ppm, 392 ppm, and 492 ppm were configured with the distribution system, and the CH_4 absorption spectrum was obtained at the pressure of 0.5 atm and the room temperature of 23.4 °C, as shown in Figure 3a. Figure 3b illustrates the linear fitting relationship between the integrated absorption area of CH_4 and its concentration. A strong linear correlation was observed, with a correlation coefficient of 99.98%, indicating a high degree of accuracy in the measurement. This result demonstrates that the system can be effectively calibrated by inverting the integrated absorption area to determine the corresponding gas concentration. The precise linearity of this relationship further suggests that the system is reliable for quantitative CH_4 detection across a wide concentration range.

3.3. Savitzky–Golay Smoothing

To verify the feasibility of Savitzky–Golay (S-G) filtering for spectral denoising, a simulation of a methane (CH_4) weak absorption line was conducted using MATLAB R2024b. First, the direct absorption spectrum (DAS) signal was superimposed with interference noise and white noise to generate a noisy signal. The noisy signal is shown in Figure 4a. To evaluate the denoising performance, signal-to-noise ratio (SNR), root mean square error (RMSE), and correlation coefficient (CC) were used as metrics to assess the correlation between the noisy signal and the denoised signal. These indicators were further analyzed to study the effectiveness of the filtering process.

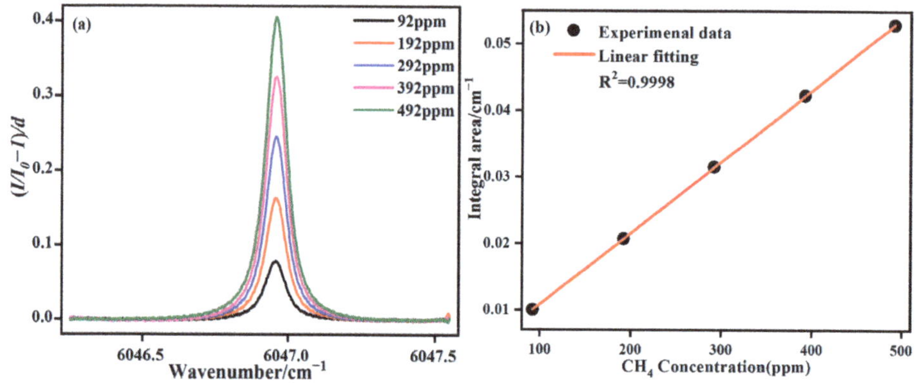

Figure 3. (**a**) The absorption coefficient spectra of methane at different concentrations. (**b**) The integral area curve of methane at different concentrations.

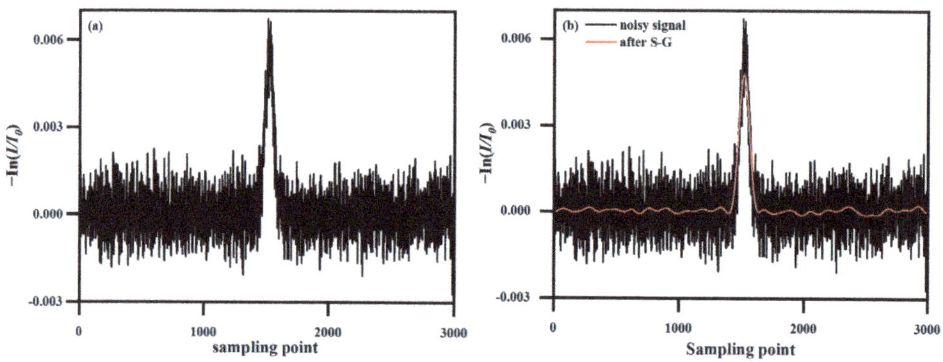

Figure 4. (**a**) Graph of the noisy signal. (**b**) Comparison with the original signal after S-G denoising.

For the noisy signal, S-G filtering was applied. Given that the SNR was close to 1, and the noise level was high, a larger window size was selected to enhance the denoising effect. Since methane detection signals are typically continuous and smooth, and a second-order polynomial (i.e., parabola) can effectively fit the smooth variations in local regions, a quadratic polynomial was chosen for fitting. The comparison between the filtered signal and the original signal is shown in Figure 4b. The calculated RMSE between the two signals was 0.0067%, SNR was 20.95, and CC was 99.5979%, demonstrating that the S-G filter effectively reduces noise.

Then, to further enhance the signal-to-noise ratio (SNR) of methane absorption signals and achieve a smoother absorption spectrum, the collected spectral data were post-processed using the Savitzky–Golay (S-G) convolution smoothing algorithm. The S-G algorithm operates by performing discrete processing, which involves weighted averaging within a sliding window. The weighting coefficients are determined by fitting a least squares polynomial within the window, enabling effective noise reduction while preserving the higher-order derivative characteristics of the data. In this study, to optimize the parameters of the Savitzky–Golay algorithm, a second-order polynomial was employed for fitting. Five different window widths (20, 40, 60, 40, 80, and 100) were evaluated on a methane signal with a concentration of 92 ppm, as illustrated in Figure 5. The results indicated that as the window width increased, the smoothing effect on the spectral line became more pronounced. However, an excessively large window width led to the loss of key spectral features, which are critical for accurate detection and quantification. To balance the trade-off between effective noise reduction and the preservation of important spectral details, a window width of 60 was selected as optimal for processing the spectral data using the S-G algorithm. The results show that the smoothing effect on the spectral lines becomes more and more obvious as the window width increases. However, an excessively large window width led to the loss of key spectral features, which are critical for accurate detection and quantification. In order to strike a balance between effective noise reduction and retention of important spectral detailing, an optimal window width of 60 was selected as optimal for processing the spectral data using the S-G algorithm to avoid signal distortion as the signal peaks flatten out at window widths of 80 and 100, and a noticeable distortion occurs.

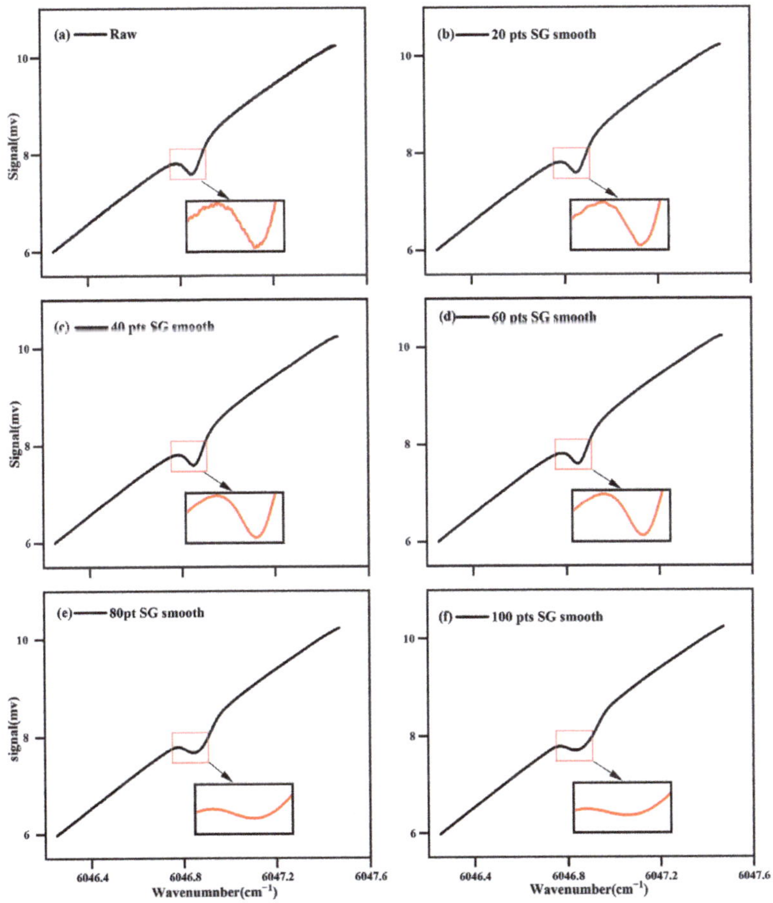

Figure 5. (**a**) Original absorption spectrum data. (**b–f**) The results of Savitzky–Golay smoothing with different window widths ranging from 20 to 100.

3.4. *Performance Evaluation*

To evaluate the performance of the CH_4 sensor, we analyze both the integrated area of CH_4 and the standard deviation of the fitting residuals at a concentration of 92 ppm, as illustrated in Figure 6. In Figure 6a, the original data yields a signal-to-noise ratio (SNR) of 93.41, and the detection accuracy of the device is 0.98 ppm, indicating the baseline sensor performance under unprocessed conditions. In contrast, Figure 6b shows the results after applying the Savitzky–Golay smoothing filter, which significantly enhances the SNR to 172.18. The application of this smoothing algorithm notably improves the signal clarity by reducing high-frequency noise, resulting in a cleaner and more accurate signal representation. Furthermore, the maximum SNR achieved with the Savitzky–Golay smoothing is approximately 1.84 times higher than the SNR obtained without the smoothing filter. This substantial improvement in SNR directly translates to enhanced sensor sensitivity, allowing for the detection of even lower concentrations. As a result, the detection accuracy of CH_4 is reduced to 0.53 ppm, demonstrating the effectiveness of the S-G smoothing filter in significantly improving the sensor performance.

To further verify the effectiveness of the S-G filter smoothing denoising algorithm, it is compared with two traditional noise reduction methods, Kalman filter noise reduction and moving average noise reduction. The direct absorption signal obtained after denoising is compared with the fitted signal curve, and the noise reduction results for each algorithm are shown in Figure 6c,d. The parameters derived from the comparison of the denoising performance of the various algorithms are summarized in Table 1. The results demonstrate that, compared to the other two methods, the root mean square error (RMSE) of the direct absorption curves obtained using the S-G filter smoothing algorithm is significantly reduced by 0.000453, the signal-to-noise ratio (SNR) is notably increased by 172.18, and the correlation is greatly enhanced, with a correlation coefficient (CC) of 99.9533%. Consequently, the direct absorption signal curve processed with the S-G filter smoothing denoising method is much closer to the fitted curve. This method shows clear noise reduction advantages, effectively restoring the peak position of the direct absorption signal while preserving its authenticity.

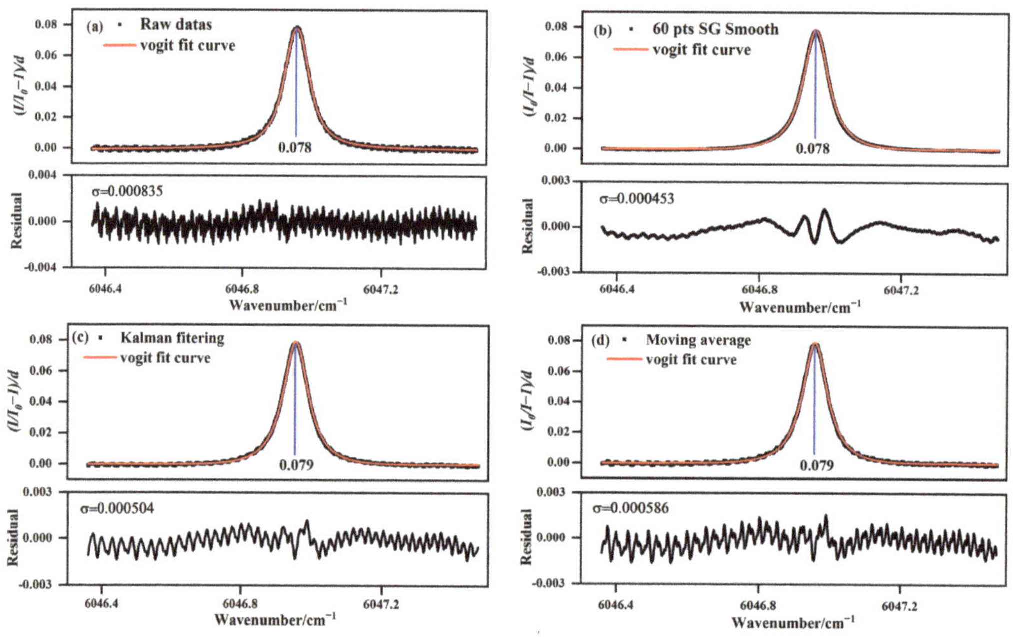

Figure 6. (**a**) The original signal. (**b**) The S-G smoothing signal. (**c**) Kalman filtering signal. (**d**) Moving average signal.

Table 1. Comparison of denoising effect.

Denoising Method	RMSE	SNR	CC/%
Raw data	8.35×10^{-4}	93.41	99.7736
60 pts SG Smooth	4.53×10^{-4}	172.18	99.9533
Kalaman filtering	5.04×10^{-4}	156.74	99.9014
Moving average	5.86×10^{-4}	134.81	99.8537

4. Conclusions

This paper presents an optical detection scheme for methane based on tunable diode laser absorption spectroscopy. Due to the degradation of signal quality from external noise, the Savitzky–Golay convolution smoothing algorithm was applied to the spectral data as a secondary processing step. This significantly enhanced the signal-to-noise ratio (SNR), improving the accuracy and reliability of the measurements. After applying the S-G smoothing, the SNR increased from an initial value of 93.41 to 172.18, which resulted in a CH_4 detection accuracy of 0.53 ppm. These results demonstrate the effectiveness of the system in detecting low-concentration methane, underscoring its potential for trace methane analysis in both environmental monitoring and industrial applications.

Author Contributions: Conceptualization, X.T., S.C. and G.C.; methodology, X.T.; software, S.C.; validation, J.Y., X.C. and S.C.; formal analysis, T.M. and S.C.; writing—original draft preparation, S.C.; writing—review and editing, X.T.; supervision, X.T., S.C. and G.C.; project administration, X.T. and G.C.; funding acquisition, X.T. and G.C. All authors have read and agreed to the published version of the manuscript.

Funding: This research was funded by the Medical Special Cultivation Project of Anhui University of Science and Technology, grant number YZ2023H1B010, and the National Natural Science Foundation of China, grant number 62405006, 62105005.

Institutional Review Board Statement: Not applicable.

Informed Consent Statement: Not applicable.

Data Availability Statement: Data are contained within the article.

Conflicts of Interest: The authors declare no conflicts of interest.

References

1. Wilson, C. Untangling variations in the global methane budget. *Commun. Earth Environ.* **2023**, *4*, 318. [CrossRef]
2. Panagiotis, S.; Giannis, P.; Michalis, V. Remote Operation of an Open-Path, Laser-Based Instrument for Atmospheric CO_2 and CH_4 Monitoring. *Photonics* **2023**, *10*, 386. [CrossRef]
3. Srivastava, R.; Kedia, S.; Rajesh, T.A. Contribution of natural and anthropogenic aerosols to optical properties and radiative effects over an urban location. *Environ. Res. Lett.* **2012**, *7*, 034028.
4. Maazallahi, H.; Delre, A.; Scheutz, C.; Fredenslund, A.M.; Schwietzke, S.; van der Gon, H.D.; Röckmann, T. Intercomparison of detection and quantification methods for methane emissions from the natural gas distribution network in Hamburg, Germany. *Atmos. Meas. Tech.* **2023**, *16*, 5051–5073. [CrossRef]
5. Chen, J.Y.; Cui, P.X.; Zhou, C.C.; Yu, X.Y.; Wu, H.H.; Jia, L.Q.; Zhou, M.; Zhang, H.J.; Teng, G.E.; Cheng, S.; et al. Detection of CO_2 and CH_4 Concentrations on a Beijing Urban Road Using Vehicle-Mounted Tunable Diode Laser Absorption Spectroscopy. *Photonics* **2023**, *10*, 938. [CrossRef]
6. Wang, R.F.; Huang, T.X.; Mei, J.X.; Wang, G.S.; Liu, K.; Chen, W.D.; Gao, X.M. Pressure sensing with two-color laser absorption spectroscopy for combustion diagnostics. *Opt. Lett.* **2024**, *49*, 1033–1036. [CrossRef] [PubMed]
7. Shi, Y.P.; Hu, Z.; Niu, M.S.; Li, T.H.; Li, H.; Liu, H.Y.; Li, X.X. High-sensitive double incidence multi-pass cell for trace gas detection based on TDLAS. *Sens. Actuators B Chem.* **2024**, *412*, 135829. [CrossRef]
8. Masiyano, D.; Hodgkinson, J.; Schilt, S.; Tatam, R.P. Self-mixing interference effects in tunable diode laser absorption spectroscopy. *Appl. Phys. B* **2009**, *96*, 863–874. [CrossRef]

9. Hartmann, A.; Strzoda, R.; Schrobenhauser, R.; Weigel, R. Ultra-compact TDLAS humidity measurement cell with advanced signal processing. *Appl. Phys. B* **2013**, *115*, 263–268. [CrossRef]
10. Li, C.L.; Guo, X.Q.; Ji, W.H.; Wei, J.L.; Qiu, X.B.; Ma, W.G. Etalon fringe removal of tunable diode laser multi-pass spectroscopy by wavelet transforms. *Opt. Quant. Electron.* **2018**, *50*, 275.1–275.11. [CrossRef]
11. Guo, X.Q.; Qiu, X.B.; Ji, W.H.; Shao, L.G.; Liu, S.P.; Li, C.L.; Ma, W.G. Minimization of Interference Fringes in Tunable Diode Laser A bs orption Spectrum Based on Empirical Mode Decomposition. *Laser Optoelectron. Prog.* **2018**, *55*, 113001.
12. Dombi, J.; Dineva, A. Adaptive Savitzky-Golay filtering and its applications. *Int. J. Adv. Intell.* **2020**, *16*, 145–156. [CrossRef]
13. Fábián, G. Generalized Savitzky–Golay filter for smoothing triangular meshes. *Comput. Aided Geom. Des.* **2023**, *100*, 102167. [CrossRef]
14. Li, J.S.; Deng, H.; Li, P.F.; Yu, B.L. Real-time infrared gas detection based on an adaptive Savitzky–Golay algorithm. *Appl. Phys. B* **2015**, *120*, 207–216. [CrossRef]
15. Tong, C.; Sima, C.; Chen, M.Q.; Zhang, X.H.; Li, T.L.; Ai, Y.; Lu, P. Laser Linewidth Analysis and Filtering/Fitting Algorithms for Improved TDLAS-Based Optical Gas Sensor. *Sensors* **2023**, *23*, 5130. [CrossRef] [PubMed]
16. Yuan, Z.H.; Huang, Y.B.; Zhong, P.; Lu, X.J.; Huang, J.; Zhang, L.L.; Qi, G.; Cao, Z.S. Design and Study of Absorption Spectrum Measurement Device with V-Shaped Off-Axis Integrated Cavity. *Chin. J. Lasers* **2023**, *50*, 207.
17. Ji, J.L.; Huang, Y.J.; Pi, M.Q.; Zhao, H.; Peng, Z.H.; Li, C.G.; Wang, Q.; Zhang, Y.; Wang, Y.D.; Zheng, C.T. Performance improvement of on-chip mid-infrared waveguide methane sensor using wavelet denoising and Savitzky-Golay filtering. *Infrared Phys. Technol.* **2022**, *127*, 04469. [CrossRef]
18. Tian, X.; Cao, Y.; Chen, J.J.; Liu, K.; Wang, G.S.; Gao, X.M. Hydrogen sulphide detection using near-infrared diode laser and compact dense-pattern multipass cell. *Chin. Phys. B* **2019**, *28*, 063301. [CrossRef]
19. Kochanov, R.V.; Gordon, I.E.; Rothman, L.S.; Shine, K.P.; Sharpe, S.W.; Johnson, T.J.; Wallington, T.J.; Harrison, J.J.; Bernath, P.F.; Birk, M.; et al. Infrared absorption cross-sections in HITRAN2016 and beyond: Expansion for climate, environment, and atmospheric applications. *J. Quant. Spectrosc. Radiat. Transfer.* **2019**, *238*, 106708. [CrossRef]
20. Liu, K.; Wang, L.; Tan, T.; Wang, G.S.; Zhang, W.J.; Chen, W.D.; Gao, X.M. Highly sensitive detection of methane by near-infrared laser absorption spectroscopy using a compact dense-pattern multipass cell. *Sens. Actuators B Chem.* **2015**, *220*, 1000–1005. [CrossRef]

photonics

MDPI

Article

A Multi-Deformable-Mirror 500 Hz Adaptive Optical System for Atmospheric Turbulence Simulation, Real-Time Reconstruction, and Wavefront Correction Using Bimorph and Tip-Tilt Correctors

Ilya Galaktionov [1,2,3,*] and Vladimir Toporovsky [1,2,3]

1 Moscow Technical University of Communications and Informatics, Aviamotornaya Str., Bld. 8a, Moscow 111024, Russia; toporovskii_v@mail.ru
2 Moscow Polytechnic University, Bolshaya Semenovskaya Str., Bld. 38, Moscow 107023, Russia
3 Sadovsky Institute of Geosphere Dynamics RAS, Leninskiy Avenue 38, Bld. 1, Moscow 119334, Russia
* Correspondence: ilya.galaktionoff@gmail.com

Abstract: Atmospheric turbulence introduces distortions to the wavefront of propagating optical radiation. It causes image resolution degradation in astronomical telescopes and significantly reduces the power density of radiation on the target in focusing applications. The impact of turbulence fluctuations on the wavefront can be investigated under laboratory conditions using either a fan heater (roughly tuned), a phase plate, or a deformable mirror (finely tuned) as a turbulence-generation device and a wavefront sensor as a wavefront-distortion measurement device. We designed and developed a software simulator and an experimental setup for the reconstruction of atmospheric turbulence-phase fluctuations as well as an adaptive optical system for the compensation of induced aberrations. Both systems use two 60 mm, 92-channel, bimorph deformable mirrors and two tip-tilt correctors. The wavefront is measured using a high-speed Shack–Hartmann wavefront sensor based on an industrial CMOS camera. The system was able to achieve a 500 Hz correction frame rate, and the amplitude of aberrations decreased from 2.6 μm to 0.3 μm during the correction procedure. The use of the tip-tilt corrector allowed a decrease in the focal spot centroid jitter range of 2–3 times from ±26.5 μm and ±24 μm up to ±11.5 μm and ±5.5 μm.

Keywords: atmospheric turbulence; Kolmogorov spectra; turbulence simulation; wavefront correction; bimorph deformable mirror; stacked-actuator deformable mirror; Shack–Hartmann wavefront sensor; adaptive optics

1. Introduction

As is known, radiation that propagates through an atmosphere is distorted due to turbulent refractive index fluctuations. This can limit telescopes' resolution as well as decrease laser radiation coherence [1–9]. Atmospheric turbulence also affects the wireless transmission of information and energy by means of optical radiation [10–13]. In particular, it limits optical communication channels in free space [14], wireless optical communication [15,16], and low-Earth-orbit satellites [11]. It also limits the ability to destroy unmanned aerial vehicles or space debris [17–19], to create a beam of the desired shape [20,21], to focus a beam inside the aperture for laser communication tasks [22], to increase the radiation power density on a target [23,24] (i.e., laser cutting) and to improve the accuracy of beam positioning for the optical recording of information.

The main reasons for the low efficiency of systems for wireless signal transmission by means of optical radiation are the diffraction of a light beam, radiation scattering via an atmospheric aerosol [25–27], and the influence of atmospheric turbulence [28–40]. The problem of wavefront degradation due to the atmospheric turbulence influence [41] has been studied for more than 50 years, but it is still relevant. And one of the most efficient ways to overcome this issue is to use adaptive optics [42,43].

The problems of increasing a laser radiation propagation range through the atmosphere using adaptive optics methods are being solved by research teams from Russia, Germany, Italy, USA, and the Netherlands [44–46]. The Air Force Maui Optical Station (AMOS) [47] in Hawaii, USA, where the ADONIS (Daylight Optical Near-Infrared System) system was built in 1993–1995, initially did not use the adaptive optics on its 1.2 m telescope. After that, the ADONIS system was moved to the 3.6 m EOAR telescope and was equipped with the adaptive optical system. In [48,49], the authors describe a vision system that uses a conventional adaptive optical system and works on the 2.5 km horizontal atmospheric path. Another vision system with adaptive optics was included; produced in the Fraunhofer Institute [50,51], it was developed to run under urban conditions. This adaptive system contains a conventional deformable mirror and works with a frequency of 800 Hz. In [52], the authors describe the use of an adaptive optical system within a 0.35 m telescope on a 20 km slant atmospheric path. In [53], the authors describe an adaptive system with a deformable mirror and a tip-tilt corrector for a 0.12 m telescope on a 3 km horizontal atmospheric path.

Most of the papers described above are devoted either to the compensation of turbulence-phase fluctuations or to turbulence generation using spatial light modulators. The novelty of our research is (1) the use of a bimorph deformable mirror in kinematic tip-tilt mounting as a turbulence-generation device and (2) the implementation of the full research cycle, which includes the simulation, generation, and correction of turbulence fluctuations. We numerically simulate phase-screen sequences using Kolmogorov theory, we experimentally reconstruct/generate a sequence of phase screens using the bimorph mirror in tip-tilt housing, and we compensate for the induced distortions using another bimorph mirror and a standalone tip-tilt corrector. The results show that the whole system works in real time—the turbulence-generation and turbulence-correction modules perform independently. The use of such a system allows for the investigation of the limitations of deformable mirrors for the correction of atmospheric turbulence fluctuations under laboratory conditions.

The sections below describe how we overcome all of these challenges, assembling and investigating a multi-deformable-mirror-adaptive optical system with turbulence-generation and wavefront-correction modules.

2. Materials and Methods

2.1. An Experimental Setup

In our previous experiments, we used a fan heater as a device to generate turbulence-phase fluctuations [54]. Prior to an experiment, we measured the wavefront distortions induced via the fan heater and confirmed that the statistics of phase fluctuations were close to Kolmogorov's statistics [55]. But the problem of the fan heater is that there is no fine tuning of the turbulence parameters—the operator cannot set the desired wind velocity, aperture diameter, or Fried radius. To overcome this issue, optical phase plates (OPPs) can be used [56,57]. An optical phase plate is a device that can simulate real atmospheric conditions in a lab setting. Another even more flexible approach is to create a completely controllable turbulence generator. To accomplish this, we implemented the model of Kolmogorov's phase screen simulation, developed the algorithm and software to reconstruct the set of simulated phase screens in real time by means of the response functions of wavefront correctors, and assembled the experimental setup with the bimorph mirror inside tip-tilt kinematic mounting that acts as a turbulence-generation device.

In [54], we used the adaptive system based on the FPGA, which allowed for the achievement of the correction frequency up to 2 kHz. FPGA processed the image bytes coming from the wavefront sensor camera "on the fly" (as they arrive from the camera), which made it possible to calculate the vector of control voltages until the whole image frame is received. This system was successfully assembled and tested, and it demonstrated rather good results—the residual error of phase fluctuations was decreased 10 times during the 2 kHz correction procedure.

However, there are a few disadvantages to systems based on the FPGA. First, a camera with an FPGA is a relatively expensive device. Second, the programming process of an FPGA processor is very time-consuming. Third, systems based on an FPGA are difficult to scale—it is impossible to rapidly implement changes if necessary. Assuming that there are also several applications in which it is necessary to compensate for phase fluctuations with a correction frequency below 1 kHz, the use of video cameras without an FPGA processor can be considered. In this case, the challenge moves to the optimization plane of the execution time of the measurements and correction algorithms as well as to the selection of an appropriate camera sensor that has the capacity to achieve the desired frame rate.

The principal optical scheme and the photo of the adaptive system for the generation and compensation of atmospheric turbulence are presented in Figure 1.

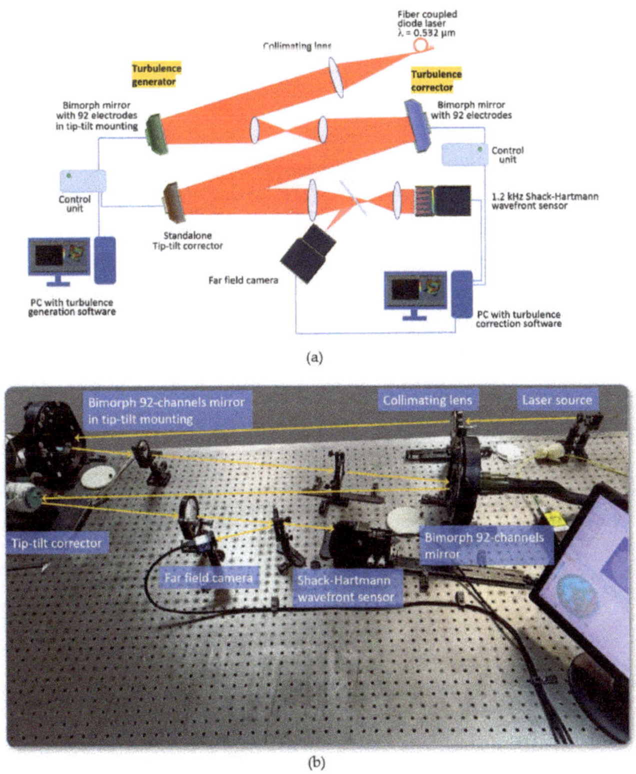

Figure 1. (**a**) Optical scheme and (**b**) photo of the complex setup with turbulence generation and turbulence correction modules.

A fiber-coupled diode laser with a wavelength of 0.532 µm was collimated with an achromatic lens. The collimated laser beam is then incident on the first 60 mm bimorph deformable mirror with 92 electrodes in the kinematic tip-tilt mounting—this construction allows us to introduce controllable wavefront distortions that, in our case, are equivalent to those induced by atmospheric turbulence. The distorted beam passes through the conjugating telescope (two lenses that conjugate the planes of two bimorph mirrors) and falls on the second 60 mm bimorph deformable mirror with 92 electrodes, which compensates for the introduced low- and high-order wavefront distortions. The beam then goes to the standalone tip-tilt corrector, passes through the matching telescope and falls on the Shack–Hartmann wavefront sensor based on a CMOS camera Baumer VCXU 13M, Stockach, Germany [58]. Then, part of the beam is reflected off the beam splitter and falls on the CMOS camera The Imaging Source DMK 23UX174, Bremen, Germany [59], which analyzes the intensity distribution of a focal spot in the far field. The sections below provide information about the optical components used in the experimental setup.

2.2. A Bimorph Deformable Mirror

The most significant turbulence-induced aberrations are tilts, defocus, and a few other low-order aberrations. Of course, wavefront aberrated by a turbulence is not limited to such aberrations. There are also high-order aberrations also, but their impact is not as high as that of low-order ones. Bimorph deformable mirrors are modal correctors, and thus they are very efficient in a reconstruction/compensation of low-order large-amplitude wavefront aberrations.

A conventional bimorph deformable mirror consists of a passive glass or silicon substrate coated with reflective coating and two piezoceramic disks glued to it [60]. A common (1st) electrode was applied to the internal piezoceramic disk, which was designed to change the curvature of the reflecting surface of the mirror. A set of control electrodes from the 2nd to the 92nd positions was applied to the external piezoceramic disk [61–64]. The piezoceramic disk can be divided into sections using laser engraving technology. The wires can be connected to the control electrodes via ultrasonic welding [65]. The reflecting coating can be dielectric, aluminum, silver, or gold [66,67], depending on the application in which the mirror is intended. The principal scheme of the bimorph mirror, the scheme of the electrodes' layout, and photos of the developed mirror are presented in Figure 2, and the mirror characteristics are given in Table 1.

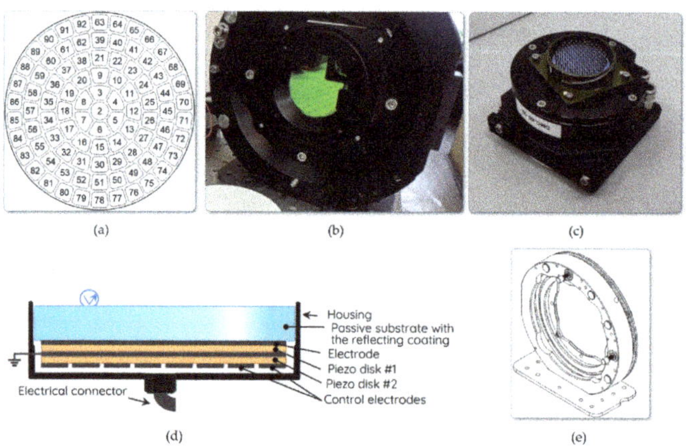

Figure 2. (**a**) Bimorph mirror electrodes layout scheme, (**b**,**c**) photo of the front and back of the manufactured bimorph mirror in tip-tilt mounting, (**d**) principal scheme of the bimorph mirror construction, (**e**) drawing of the tip-tilt kinematic mounting.

Table 1. Parameters of the bimorph deformable mirror in tip-tilt mounting.

Parameter	Value
Substrate aperture	65 mm
Clear aperture	60 mm
Substrate material	Silica
No. of PZT	2
No. of control electrodes	92
Type of actuators	PZT discs
Actuators geometry	sectorial
Maximum input voltage	−300–+500 V
Maximum element diameter for tip-tilt mounting	152 mm
Angle displacement range of tip-tilt mounting	±35 mrad

2.3. A Tip-Tilt Corrector

Most of the spectral power of the Kolmogorov spectra phase fluctuations was within the first few low-order wavefront aberrations, i.e., tip-tilt, defocus, astigmatism, and coma, according to Taylor's hypothesis [68]. Moreover, tip-tilt aberration has a greater than 40% impact on the spectral power, indicating that these aberrations should be considered. One effective device to compensate for tip-tilt aberration is a mirror with magnetostriction [69] or piezoelectric actuators [70,71].

The piezo tilt stage (two-axis motion) was based on a parallel kinematic design with a coplanar axis and moving platform. Four piezo actuators were placed at a 90° angle interval paired differential control distribution. Two pairs of differentially driven actuators provide the highest achievable angular stability over a wide temperature range. The tilt motion is achieved by two pairs of piezo actuators in a push–pull motion controlled by a bridge connection circuit.

Piezoelectric actuators typically consist of stacks of multilayer piezoelectric materials enclosed in a metal housing for preloading [72]. Applying a modulated high-voltage signal to piezoelectric ceramics causes a small deformation of the material. Compared with electromagnetic coils, piezoelectric actuators generate enormous forces in a smaller housing with a much higher frequency response. However, such actuators have a limited range of motion, exhibit hysteresis, and must have a preload mechanism to prevent the piezoelectric stack from delaminating and to overcome external forces. The combination of the high operating frequency, load, and stability of small tilt settings facilitates the use of piezoelectric actuators to drive mirrors to correct wavefront tilts. The design of the tilt corrector, which is based on piezoelectric actuators along with an electronic drive circuit, is shown in Figure 3.

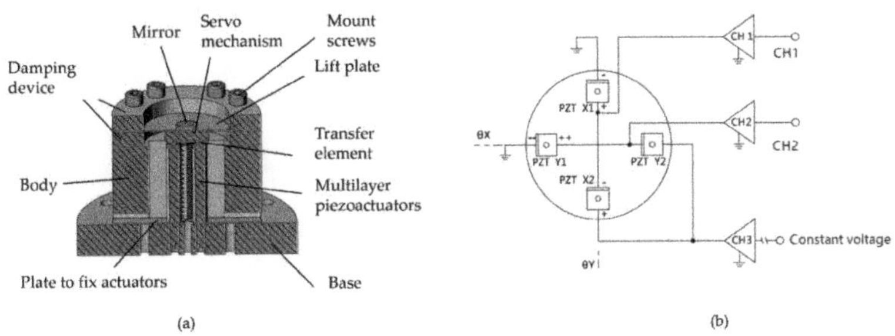

Figure 3. (**a**) Design drawing and (**b**) electronic drive circuit of the tip-tilt corrector.

The local deformation of one pair of piezo actuators forms a rotation center (pivot point) and at the same time preloads the piezo actuator. The drives operate electrically on a bridge circuit, which is supplied with a fixed voltage and controlled by an alternating voltage.

The tip-tilt corrector's linearity was calibrated and tested. The diagram in Figure 4 demonstrates the response linearity of 0.2%.

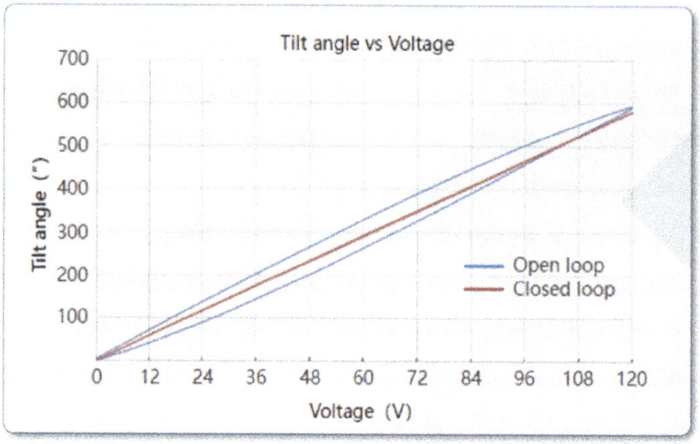

Figure 4. Dependence of the tilt angle on the voltage applied to an actuator.

The main parameters of the tip-tilt corrector are listed in Table 2.

Table 2. Parameters of the tip-tilt corrector.

Parameter	Value
Clear aperture	56 mm
Substrate material	Silica
No. of control actuators	4
Angle displacement	±0.4 mrad
Voltages range	−0–+180 V
Control frequency	More than 200 Hz
Resolution	0.02 µrad
Unloaded resonant frequency	1.5 kHz
Unloaded step time	2 ms
Electrical capacitance	7.2/axis µF
Operating temperature	−20~80 °C
Material	Aluminum
Closed-loop linearity	0.2% F.S.
Closed-loop repeatability	0.02% F.S.

2.4. A Shack–Hartmann Wavefront Sensor

In order to measure the wavefront aberrations, a Shack–Hartmann wavefront sensor [73–78] was used. A Shack–Hartmann sensor is a robust, easily calibrated, and widely used device in a large and diverse set of applications, and it is primarily used to measure wavefront aberrations of radiation passing through a turbulent or scattering atmosphere [79], biological tissues [80], etc.

The principle of a conventional Shack–Hartmann sensor can be described as follows. The wavefront of the incident light is divided into a set of subapertures using a microlens array. The microlens array is a thin, flat base with a grid of microlenses etched on it. Each microlens has (usually) a diameter of 100 to 300 µm and a focal length f of 3–8 mm. A field of focal spots is created at the focal plane of the microlens array when the radiation passes through this array (Figure 5).

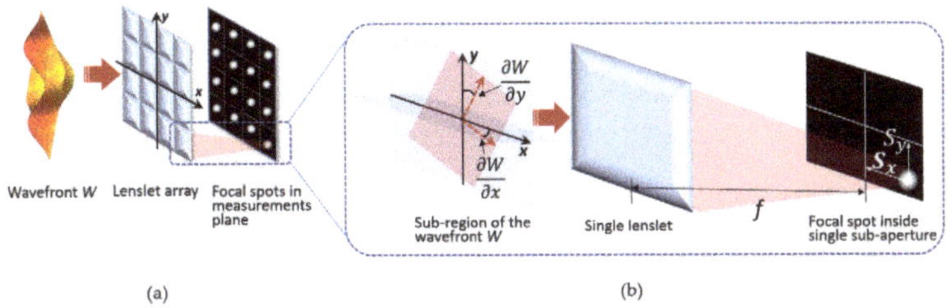

Figure 5. (**a**) Set of focal spots (also called focal spot field, Hartmannogram) formed by microlens array at the sensor plane of a Shack–Hartmann sensor; (**b**) principle of calculation of wavefront derivatives by means of measurements of focal spot displacements.

Because the diameter of each microlens is small, wavefront W is assumed to be flat and only exhibit tip-tilt aberration within a single microlens. In the case of no aberrations (i.e., a wavefront is flat and parallel to the plane of the microlens), the radiation is focused at the center of the corresponding subaperture of the sensor. If the wavefront in a microlens has a non-zero tip-tilt, then the focal spot is displaced (S_x and S_y) from the center of the subaperture in proportion to the tip-tilt value. In other words, if we measure the displacements S_x and S_y of the focal spot per X and Y axis, we can obtain the values of the partial derivatives $\partial W/\partial x$ and $\partial W/\partial y$ of the wavefront W within each subaperture:

$$\left\{ \begin{array}{c} \frac{\partial W(x,y)}{\partial x} \\ \frac{\partial W(x,y)}{\partial y} \end{array} \right\} = \frac{1}{f} \left\{ \begin{array}{c} S_x \\ S_y \end{array} \right\}, \tag{1}$$

where $\frac{\partial W(x,y)}{\partial x}$, $\frac{\partial W(x,y)}{\partial y}$ are the partial derivatives of wavefront W, f is the focal length of each lenslet, S_x, S_y are the displacements of each focal spot within a subaperture per the X and Y axis.

On the other hand, to describe and visualize a wavefront surface analytically, one can use a polynomial approximation, for example, B-splines [81] or Zernike polynomials [82–85], which are commonly used in optics. Thus, the partial wavefront derivatives $\partial W/\partial x$ and $\partial W/\partial y$ can be defined analytically using Zernike polynomials:

$$\left\{ \begin{array}{c} \frac{\partial W(x,y)}{\partial x} \\ \frac{\partial W(x,y)}{\partial y} \end{array} \right\} = \left\{ \begin{array}{c} \sum_i^N a_i \cdot \frac{\partial Z_i(x,y)}{\partial x} \\ \sum_i^N a_i \cdot \frac{\partial Z_i(x,y)}{\partial y} \end{array} \right\}, \tag{2}$$

where a_i represents a Zernike coefficient representing the aberration value, Z_i represents the ith Zernike polynomial, N represents the number of Zernike polynomials used.

The values can also be calculated from the measured displacements S_x and S_y of the focal spots on a Shack–Hartmann sensor:

$$\left\{ \begin{array}{c} \frac{\partial W(x,y)}{\partial x} \\ \frac{\partial W(x,y)}{\partial y} \end{array} \right\} = \left\{ \begin{array}{c} \sum_i^N a_i \cdot \frac{\partial Z_i(x,y)}{\partial x} \\ \sum_i^N a_i \cdot \frac{\partial Z_i(x,y)}{\partial y} \end{array} \right\} = \frac{1}{f} \left\{ \begin{array}{c} S_x \\ S_y \end{array} \right\}. \tag{3}$$

Finally, we obtain an overdetermined system of linear equations with unknown coefficients a_i. By solving the least squares problem [86], we obtain the coefficients a_i. The wavefront then be analytically described and analyzed.

For our experimental setup, we developed the Shack-Hartmann wavefront sensor based on a CMOS industrial camera Baumer VCXU 13M [58]. This camera operates at 222 fps at full resolution of 1280 $\times$ 1024 pixels (1/2"). By adjusting the region of interest (ROI) and setting it to 480 $\times$ 480 pixels, we were able to increase the frame rate up to 1200 fps, which in turn allowed us to achieve the wavefront correction frequency up to 600 Hz (up to 500 Hz stable) in a closed loop. A photo of the developed wavefront sensor and its parameters is presented in Figure 6.

(a)

sensor information	
sensor	ON Semiconductor PYTHON1300
resolution	1280 × 1024 px
exposure time	0.02 ... 1000 ms
pixel size	4.8 × 4.8 µm
shutter type	Global shutter
sensor type	1/2" CMOS
acquisition formats	
image formats, interface frame rate max.	Full Frame, 1280 × 1024 px, max. 222 fps Binning 2×2, 640 × 512 px, max. 222 fps Binning 2×1, 640 × 1024 px, max. 222 fps Binning 1×2, 1280 × 512 px, max. 222 fps
pixel formats	Mono8 Mono10

(b)

Figure 6. (**a**) Photo and (**b**) parameters of the Shack–Hartmann wavefront sensor based on a Baumer VCXU 13M CMOS camera.

2.5. Algorithm of Phase Screens Simulation and Reconstruction

As is known [87], the simplest and most reliable model of radiation propagation through a turbulent atmosphere is a thin phase-screen model. The aberrations of an optical wave that passes through a set of thin phase screens are similar to the fluctuations of the light field in a continuous randomly inhomogeneous medium. The thin phase screen closely reproduces the influence of large-scale atmospheric inhomogeneities on the characteristics of the light field [88]. The phase screen approach presents a good approximation, and in most cases, it can reproduce the effect of turbulence on the wavefront with acceptable accuracy.

In order to simulate a set of phase screens, we applied the Fast Fourier transform to the Kolmogorov spectrum of the phase fluctuations [89]:

$$p(u,v,t+\Delta t) = \iint_{-\infty}^{\infty} \sqrt{K(x,y)}\cdot f(x,y,t+\Delta t)\cdot e^{i\cdot\sqrt{x^2+y^2}\cdot V\cdot\Delta t}dxdy, \qquad (4)$$

where $p(u,v,t+\Delta t)$ is the phase screen at $t+\Delta t$, (x,y) is a spectrum point, (u,v) is a phase screen point, V is the wind velocity, m/s, t is the moment of the previous phase screen generation, $t+\Delta t$ is the time moment of the new phase screen generation, Δt is the time interval between two phase screens, and $K(x,y)$ is the spectrum of the phase fluctuations.

To calculate the spectrum of the phase fluctuations $K(x,y)$, the following formula was used [90]:

$$K(x,y) = 0.023\cdot(\frac{2D}{r_0})^{\frac{5}{3}}\cdot(x^2+y^2)^{-\frac{11}{3}}, \qquad (5)$$

where D is the telescope receiving aperture, r_0 is a Fried radius, and $f(x,y,t+\Delta t)$ is a function defined using the following formula:

$$f(x,y,t+\Delta t) = p{\cdot}f(x,y,t) + \sqrt{1-p^2}{\cdot}e^{i\cdot scale=0.9\varphi(x,y,t)}, \tag{6}$$

where $p = e^{\frac{-\Delta t}{\tau}}$, τ is the atmosphere coherence time, and $scale = 0.9\varphi(x,y,t)$ is a random delta-correlated value in the range of $[0; 2\pi]$.

When $t = 0$, the function f is expressed as shown below:

$$f(x,y,t=0) = e^{i\cdot scale=0.9\varphi(x,y,t=0)}. \tag{7}$$

The calculated phase values were normalized in accordance with the relation $\frac{D}{r_0}$ [46] using the following phase structure function:

$$DN = 6.88{\cdot}\left(\frac{\sqrt{x^2+y^2}}{r_0}\right)^{\frac{5}{3}}. \tag{8}$$

Once the phase screen is generated, it can be approximated using Zernike polynomials. In this way, the simulated and approximated phase maps can be numerically compared, i.e., by calculating the point-to-point phase difference. The resultant phase maps are presented in Figure 7.

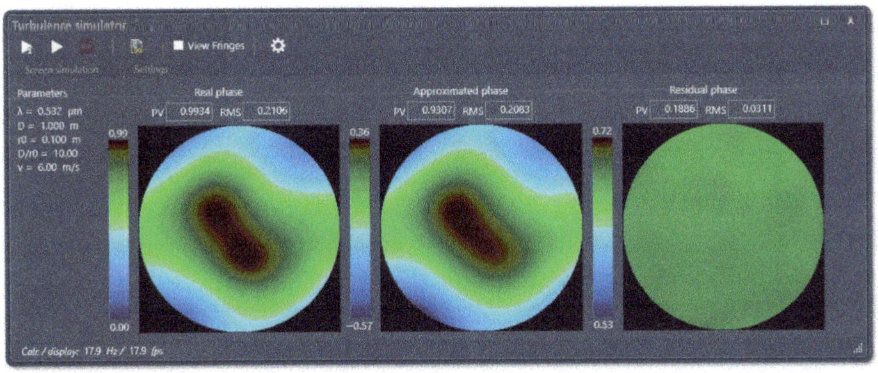

Figure 7. Screenshot of the desktop software tool for a phase screen simulation: from left to right—a simulated phase screen according to Kolmogorov spectra, a phase screen approximated by Zernike polynomials, and a phase difference map.

After implementation of the phase screen simulation model, we conducted an analysis of the energy distribution of Zernike polynomials within the simulated turbulent wavefront. The diagram in Figure 8 illustrates the influence of each Zernike mode on the resulting wavefront based on a substantial set of simulated phase screens. Additionally, we analyzed the energy distribution during continuous wavefront correction both with and without tip-tilt correction. The results indicate a strong alignment between the observed energy distribution and the theoretical predictions. The corresponding bar diagram is presented in Figure 8.

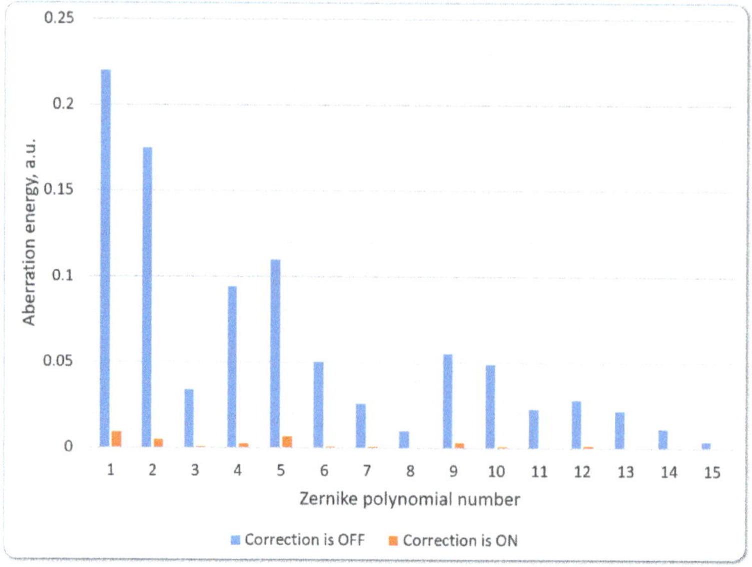

Figure 8. Energy distribution over Zernike polynomials in the simulated turbulent phase screens.

Once the set of phase screens was simulated, the procedure of phase screens approximation using the response functions of the bimorph mirror [91] started. The response function for a single electrode of the deformable mirror describes how the mirror's surface deforms in response to the voltage applied to that specific electrode [92]. Response functions are measured using a Shack–Hartmann sensor.

To reconstruct the phase screen with the mirror, the following procedure was used:

1. Calculation of the wavefront derivatives in each subaperture of the wavefront sensor based on the known Zernike approximation of the simulated phase screen, as previously described above and illustrated in Figure 7.
2. Calculation of the displacements of the focal spots corresponding to the wavefront derivatives calculated in Step 1.
3. By knowing the focal spot shifts related to the mirror's response functions and the focal spot displacements needed to reproduce the wavefront, we were able to solve the overdetermined system of linear equations using the least squares method [93]. This allowed us to calculate the voltage vector that needed to be applied to the mirror electrodes [94].

This procedure was repeated for each simulated phase screen, resulting in a set of voltage vectors. Each vector corresponds to a particular phase screen, meaning that applying this vector of voltages to the mirror electrodes should generate a wavefront that aligns with the simulated phase screen.

3. Results

The principal steps of the experiment are outlined as follows:

1. Simulation of phase screens utilizing the Fast Fourier transform to model the Kolmogorov spectrum of phase fluctuations [1,95].
2. Determination of the voltage set required for the reconstruction of the phase screens [96].
3. Real-time reconstruction of the simulated phase screens employing the bimorph mirror mounted in a tip-tilt mounting.

4. Measurement of wavefront distortions in real time using the Shack–Hartmann wavefront sensor along with an analysis of the intensity distribution of the focal spot in the far field.
5. Real-time computation of the correction voltages necessary for both the bimorph deformable mirror and the tip-tilt corrector.
6. Real-time compensation of the reconstructed phase screens.

The primary parameter characterizing the intensity of atmospheric turbulence is the refractive index structure parameter C_n^2 [95]; specifically, a larger C_n^2 value indicates stronger turbulence. This parameter can range from $10^{-17}\text{m}^{-\frac{2}{3}}$ for weak turbulence to $10^{-12}\text{m}^{-\frac{2}{3}}$ for very strong atmospheric turbulence. For instance, in Hefei, China, C_n^2 has been observed to vary between $6.69 \times 10^{-16}\text{m}^{-\frac{2}{3}}$ and $9.87 \times 10^{-14}\text{m}^{-\frac{2}{3}}$ during summer measurements along a horizontal atmospheric path of 1 km [97]. In maritime conditions, C_n^2 is approximately $10^{-15}\text{m}^{-\frac{2}{3}}$ for a 10 km path with a coherence radius r_0 of 3.8 cm at a laser wavelength $\lambda = 0.85$ μm [98].

For ground-to-space communications between the International Space Station and the Optical Communications Telescope Laboratory (OCTL) in Wrightwood, California, a coherence radius of $r_0 = 4.5$ cm was experimentally determined for a 1200 km path at a zenith angle 75° with input telescope apertures ranging from 10 cm to 100 cm [99]. In terrestrial atmospheric turbulence, C_n^2 is approximately $10^{-12}\text{m}^{-\frac{2}{3}}$ for a 1 km path under conditions where the wind velocity is 10 m/s and the telescope aperture is 20 cm [100,101]. In desert environments, C_n^2 measures around $10^{-13.2}\text{m}^{-\frac{2}{3}}$ (with an average wind velocity of 6 m/s) along a 1.2 km path at Edward Air Force Base in the Mojave Desert, California, USA [102].

For the atmospheric turbulence measured by our team at the 1.2 km intra-city link, the coherence radius was equal to $r_0 = 1.6$ cm for a laser wavelength $\lambda = 0.532$ μm and $r_0 = 2.65$ cm for a laser wavelength $\lambda = 0.81$ μm with wind velocities ranging from 5 to 10 m/s for a receiving aperture of D = 140 mm.

Furthermore, C_n^2 can be derived from the known receiving aperture diameter D and the Fried radius r_0. In this study, we set $\frac{D}{r_0} = 20$ with a wavelength λ = 0.532 μm and a wind velocity $v = 6$ m/s. Consequently, C_n^2 varies approximately from $3.6 \times 10^{-14}\text{m}^{-\frac{2}{3}}$ to $2.2 \times 10^{-13}\text{m}^{-\frac{2}{3}}$ for path lengths ranging from 500 m to 3 km.

A comprehensive set of over 1000 phase screens was generated based on the specified parameters. Selected sequential phase screens from this dataset are illustrated in Figure 9.

Several parameters serve as input data for the phase screen simulator, which is based on Kolmogorov's theory, including aperture diameter, Fried radius, and wind velocity. Given that the model operates in a quasi-static manner—generating only one single phase screen at a time—while turbulence is inherently a volumetric phenomenon (wherein the distortions imposed on the wavefront cannot be accurately represented by a single phase screen), it is necessary to simulate a series of turbulent phase screens. The parameter that governs this temporal effect within the model is wind velocity. The choice of wind velocity influences the degree of difference between sequentially simulated phase screens; as wind velocity increases, the variation between these screens becomes more pronounced, leading to a less similar appearance from one screen to the next. This behavior is illustrated in Figure 9, where the phase appears to "drift" gradually from one screen to another.

Once the turbulence generation module is running, we can simultaneously activate the turbulence correction module. Both modules feature adjustable frequencies, allowing independent control over the generation frequency of turbulence phase screens and the correction frequency. Given that the correction frequency is constrained to 500 Hz by the wavefront sensor, the maximum frequency of turbulence that can be effectively corrected is below 30 Hz [103].

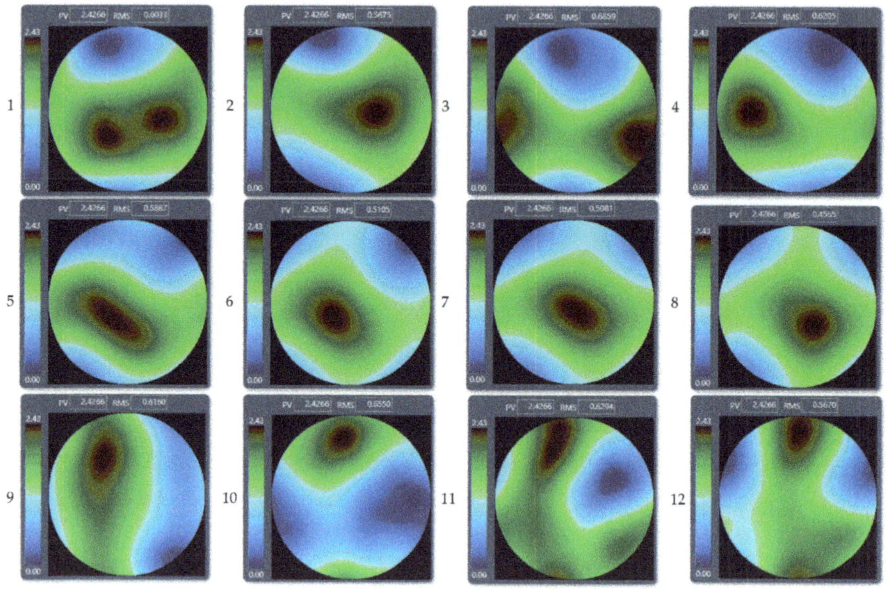

Figure 9. A selection of phase maps from the generated set of phase screens, characterized by a ratio of $\frac{D}{r_0} = 20$. The term PV denotes peak to valley, which represents the wavefront amplitude in microns, while RMS refers to the root mean square error in microns. The numbers 1 to 12 correspond to the sequentially generated phase screens.

During the continuous correction of induced wavefront distortions, the amplitude of the residual wavefront aberrations ranged from 0.25 to 0.35 μm, whereas the initial amplitude was between 2.5 and 2.8 μm. An example of the wavefront before and after correction is presented in Figure 10.

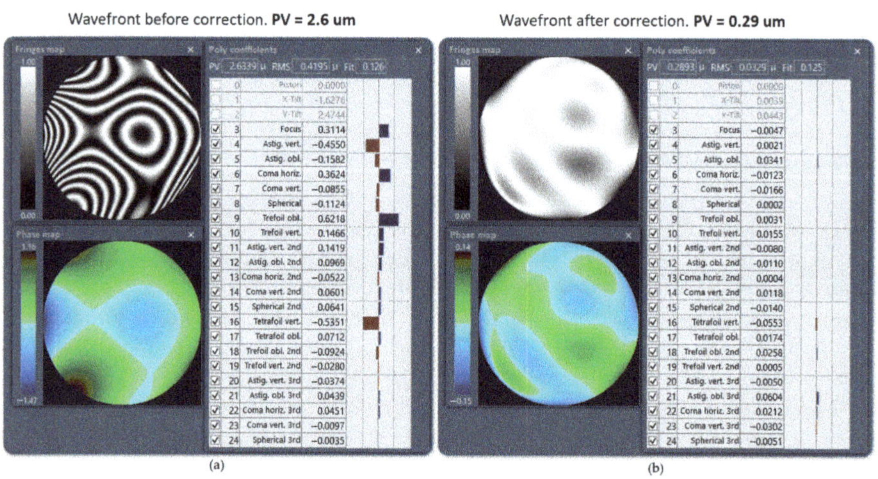

Figure 10. (**a**) The wavefront corresponded to a generated turbulence phase screen and (**b**) the wavefront after correction using a bimorph mirror in conjunction with a tip-tilt corrector.

It is important to emphasize that not only were the common Zernike coefficients—such as defocus, astigmatism, etc.—significantly reduced, but the tip and tilt components were also minimized. Specifically, the tip and tilt values were decreased from −1.62 μm and 2.47 μm to 0.004 μm and 0.04 μm, respectively, due to the use of the standalone tip-tilt corrector.

The significance of the tip-tilt corrector becomes particularly clear from the following Figure 11.

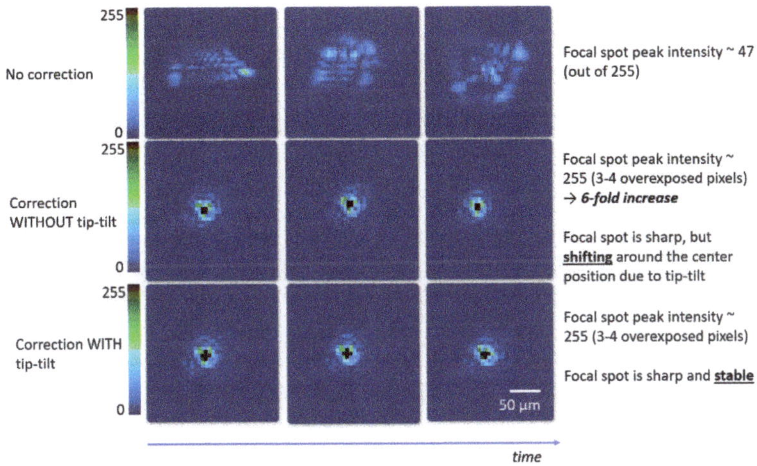

Figure 11. The far field focal spot stabilization with and without tip-tilt correction. The color palette employed is for illustrative purposes only—the row data are represented in 8-bit format (256 shades of gray).

The first row of Figure 11 contains images of the far field focal spot captured by a video camera while the turbulence generator was in operation. It is evident that the focal spot suffers significant degradation with the peak intensity value (expressed in shades of gray) measuring only 47 out of 255.

The second row of Figure 11 displays images of the focal spot when the correction module #1 was activated, utilizing the bimorph mirror to mitigate wavefront distortions while the tip-tilt corrector remained inactive. A notable enhancement in the intensity distribution of the focal spot is observed with the peak intensity increasing sixfold (reaching up to 255 with 3–4 overexposed pixels). Despite this substantial improvement, it is important to note that the focal spot remains unsteady as it shifts within the aperture. This effect is illustrated in the figure, where the focal spot is not centered within the aperture.

The third row of Figure 11 illustrates images of the focal spot when both correction modules were turned on: the bimorph mirror correcting wavefront distortions while the tip-tilt corrector corrects for tip and tilt aberrations. In this scenario, no shifts of the focal spot within the aperture were observed, resulting in a sharp and stable focal spot.

Figure 12 depicts the jitter curves of the centroid of the far field focal spot prior to correction, during correction with the bimorph mirror alone, and with both the bimorph mirror and the tip-tilt corrector operating in tandem. The frequency of the turbulent phase screen reproduction by the turbulence generator was set to 20 Hz.

(a)

(b)

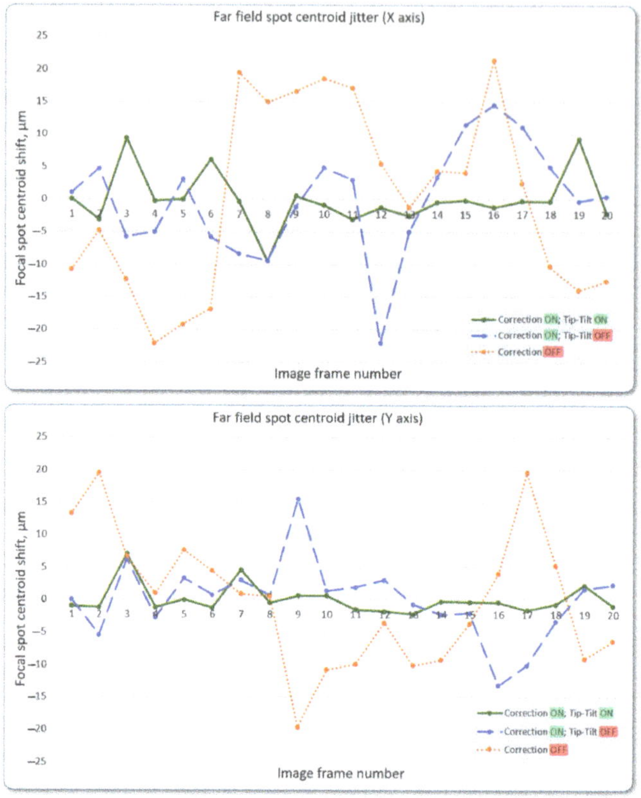

Figure 12. Jitter curves of the centroid of the far field focal spot along (**a**) the X-axis and (**b**) the Y-axis. The orange dotted curve represents the centroid displacements when the correction is disabled and the turbulence generator is active. The blue dashed curve illustrates the centroid displacements during operation of the bimorph mirror alone. The green solid curve depicts the centroid displacements when both the bimorph mirror and the tip-tilt corrector were in operation.

When the turbulence generator is operational, the amplitude of the centroid displacements of the far field focal spot was measured at 53 μm along the *X*-axis and 48 μm along the *Y*-axis. During correction with only the bimorph mirror (without the tip-tilt corrector), the amplitude of the focal spot displacements was reduced to 45 μm along *X*-axis and 35 μm along the *Y*-axis. However, when both the bimorph mirror and the tip-tilt corrector were activated, the amplitude of the focal spot displacements decreased by a factor of 2 to 4, yielding values of 23 μm along the X-axis and 11 μm along the Y-axis.

Both the charts and the numerical data indicate that the residual displacements of the focal spot centroid are ±11.5 μm along the *X*-axis and ±5.5 μm along the *Y*-axis. Assuming that the pixel size of the camera sensor is 5.86 μm, this suggests that the centroid of the focal spot shifts within a range of approximately 2 pixels during the correction procedure.

There are at least two factors that may contribute to this phenomenon. First, during the experiments, it was observed that the mounting of the standalone tip-tilt corrector requires modification or replacement, as additional vibrations were detected when the operational frequency was increased to 300 Hz during the correction mode. Second, the upper-level program managing the closed-loop system works under the Windows operating system, which must contend with the Windows messaging loop. This can introduce minor delays in the operational procedure. In general, these delays can coincide with the moments when new phase screens are generated by the turbulence generator, leading to shifts in the focal spot.

4. Conclusions

To summarize, in this research, we successfully implemented a model for simulating turbulent phase fluctuations characterized by Kolmogorov spectra. We designed and constructed an experimental setup for real-time turbulence generation as well as an apparatus for wavefront correction. Furthermore, we developed an algorithm and software capable of reconstructing a series of phase screens in real time, utilizing the response functions of a bimorph mirror mounted in a tip-tilt kinematic mounting.

The controllable turbulence generator we created allowed for the generation of phase screens with predefined parameters, including wind velocity, aperture diameter, and Fried radius, in real time. In our experiments, the parameters were set as follows: $\frac{D}{r_0}$ was 10, the wavelength was 0.532 μm, and the wind velocity was 6 m/s. The refractive index structure parameter C_n^2 ranged approximately from $3.6 \times 10^{-14} \mathrm{m}^{-\frac{2}{3}}$ to $2.2 \times 10^{-13} \mathrm{m}^{-\frac{2}{3}}$ for path lengths varying from 500 m to 3000 m. The induced wavefront aberrations exhibited a frequency of 20 Hz and amplitudes ranging from 2.4 to 2.8 μm.

The adaptive optical system functioned at a closed-loop frequency of 500 Hz, successfully reducing wavefront distortion amplitudes to as low as 0.3 μm. During the correction process utilizing both the bimorph mirror and the tip-tilt corrector, the jitter of the far field focal spot centroid was decreased by a factor of 2.5 to 4 compared to the no-correction mode, while the peak intensity of the focal spot was increased sixfold.

Author Contributions: Conceptualization, I.G.; methodology, I.G.; software, I.G.; validation, I.G. and V.T.; formal analysis, I.G.; investigation, I.G. and V.T.; resources, V.T.; data curation, V.T.; writing—original draft preparation, I.G.; writing—review and editing, I.G. and V.T.; visualization, I.G.; supervision, I.G.; project administration, I.G.; funding acquisition, I.G. and V.T. All authors have read and agreed to the published version of the manuscript.

Funding: This research received no external funding.

Institutional Review Board Statement: Not applicable.

Informed Consent Statement: Not applicable.

Data Availability Statement: Data are contained within the manuscript.

Acknowledgments: The authors would like to thank Anna Lylova for the turbulence simulation algorithm and software support. The authors also thank the laboratory of atmospheric adaptive optics of Sadovsky Institute of Geosphere Dynamics for technical support and assistance.

Conflicts of Interest: The authors declare no conflicts of interest.

References

1. Tatarsky, V.I. *Waves Propagation in Turbulent Atmosphere*; Nauka: Moscow, Russia, 1967.
2. Andrews, L.C.; Phillips, R.L. *Laser Beam Propagation Through Random Media*, 2nd ed.; SPIE Press: Bellingham, WA, USA, 2005.
3. Huang, Q.; Liu, D.; Chen, Y.; Wang, Y.; Tan, J.; Chen, W.; Liu, J.; Zhu, N. Secure free-space optical communication system based on data fragmentation multipath transmission technology. *Opt. Express* **2018**, *26*, 13536–13542. [CrossRef] [PubMed]
4. Nafria, V.; Han, X.; Djordjevic, I. Improving free-space optical communication with adaptive optics for higher order modulation. In Proceedings of the Optics and Photonics for Information Processing XIV, Online, CA, USA, 24 August–4 September 2020.
5. Vorontsov, M.; Weyrauch, T.; Carhart, G.; Beresnev, L. Adaptive Optics for Free Space Laser Communications. In *Lasers, Sources and Related Photonic Devices, OSA Technical Digest Series (CD)*; Optica Publishing Group: Washington, DC, USA, 2010; p. LSMA1.
6. Weyrauch, T.; Vorontsov, M. Free-space laser communications with adaptive optics: Atmospheric compensation experiments. *J. Optic. Comm. Rep.* **2004**, *1*, 355–379. [CrossRef]
7. Lema, G. Free space optics communication system design using iterative optimization. *J. Opt. Commun.* **2024**, *44*, s1205–s1216. [CrossRef]

8. Zhang, Y.; Wang, Y.; Deng, Y.; Du, A.; Liu, J. Design of a Free Space Optical Communication System for an Unmanned Aerial Vehicle Command and Control Link. *Photonics* **2021**, *8*, 163. [CrossRef]
9. Majumdar, A. Fundamentals of Free-Space Optical (FSO) Communication System. In *Advanced Free Space Optics (FSO)*; Series in Optical Sciences; Springer: New York, NY, USA, 2015; Volume 186.
10. Lu, M.; Bagheri, M.; James, A.P.; Phung, T. Wireless Charging Techniques for UAVs: A Review, Reconceptualization, and Extension. *IEEE Access* **2018**, *6*, 29865–29884. [CrossRef]
11. Geoffrey, A.; Landis, H. Laser beamed power—Satellite demonstration applications. In Proceedings of the 43rd International Astronautical Congress, Washington, DC, USA, 28 August–5 September 1992.
12. Vorontsov, M.; Weyrauch, T.; Carhart, G. *Lasers, Sources and Related Photonic Devices*; OSA Technical Digest Series (CD); Optica Publishing Group: San Diego, USA, 2010; p. LSMA1.
13. Galaktionov, I.; Kudryashov, A.; Sheldakova, J.; Nikitin, A. Laser beam focusing through the scattering medium by means of adaptive optics. In Proceedings of the Adaptive Optics and Wavefront Control for Biological Systems III, San Francisco, CA, USA, 28 January–2 February 2017.
14. Wang, R.; Wang, Y.; Jin, C.; Yin, X.; Wang, S.; Yang, C.; Cao, Z.; Mu, Q.; Gao, S.; Xuan, L. Demonstration of horizontal free-space laser communication with the effect of the bandwidth of adaptive optics system. *Opt. Commun.* **2018**, *431*, 167–173. [CrossRef]
15. Chittoor, P.K.; Chokkalingam, B.; Mihet-Popa, L. A Review on UAV Wireless Charging: Fundamentals, Applications, Charging Techniques and Standards. *IEEE Access* **2021**, *9*, 69235–69266. [CrossRef]
16. Wang, C.; Ma, Z. Design of wireless power transfer device for UAV. In Proceedings of the 2016 IEEE International Conference on Mechatronics and Automation, Harbin, China, 7–10 August 2016; pp. 2449–2454.
17. Bennet, F.; Conan, R.; D'Orgeville, C.; Dawson, M.; Paulin, N.; Price, I.; Rigaut, F.; Ritchie, I.; Smith, C.; Uhlendorf, K. Adaptive optics for laser space debris removal. In Proceedings of the Adaptive Optics Systems III, Amsterdam, The Netherlands, 1–6 July 2012.
18. Phipps, C.R.; Baker, K.L.; Libby, S.B.; Liedahl, D.A.; Olivier, S.S.; Pleasance, L.D.; Rubenchik, A.; Trebes, J.E.; George, E.V.; Marcovici, B.; et al. Removing orbital debris with lasers. *Adv. Space Res.* **2012**, *49*, 1283–1300. [CrossRef]
19. Shen, S.; Jin, X.; Hao, C. Cleaning space debris with a space-based laser system. *Chin. J. Aeronaut.* **2014**, *27*, 805–811. [CrossRef]
20. Sheldakova, J.; Galaktionov, I.; Nikitin, A.; Rukosuev, A.; Kudryashov, A. LC phase modulator vs deformable mirror for laser beam shaping: What is better? In Proceedings of the Laser Beam Shaping XVIII, San Diego, CA, USA, 19–23 August 2018.
21. Galaktionov, I.; Kudryashov, A.; Sheldakova, J.; Nikitin, A. Laser beam focusing through the scattering medium using bimorph deformable mirror and spatial light modulator. *Proc. SPIE* **2019**, *11135*, 111350B.
22. Galaktionov, I.; Nikitin, A.; Sheldakova, J.; Toporovsky, V.; Kudryashov, A. Focusing of a Laser Beam Passed through a Moderately Scattering Medium Using Phase-Only Spatial Light Modulator. *Photonics* **2022**, *9*, 296. [CrossRef]
23. Soloviev, A.; Kotov, A.; Martyanov, M.; Perevalov, S.; Zemskov, R.; Starodubtsev, M.; Alexandrov, A.; Galaktionov, I.; Samarkin, V.; Kudryashov, A.; et al. Improving focusability of post-compressed PW laser pulses using a deformable mirror. *Opt. Express* **2022**, *30*, 40584–40591. [CrossRef] [PubMed]
24. Kotov, A.; Perevalov, S.; Starodubtsev, M.; Zemskov, R.; Alexandrov, A.; Galaktionov, I.; Kudryashov, A.; Samarkin, V.; Soloviev, A. Adaptive system for correcting optical aberrations of high-power lasers with dynamic determination of the reference wavefront. *Quantum Electron.* **2021**, *51*, 593–596. [CrossRef]
25. Galaktionov, I.; Kudryashov, A.; Sheldakova, J.; Nikitin, A. The use of modified hill-climbing algorithm for laser beam focusing through the turbid medium. In Proceedings of the Laser Resonators, Microresonators, and Beam Control XIX, San Francisco, CA, USA, 28 January–2 February 2017.
26. Kalinskaya, D.V.; Papkova, A.S. Optical characteristics of atmospheric aerosol from satellite and photometric measurements at the dust transfers dates. In Proceedings of the 26th International Symposium on Atmospheric and Ocean Optics, Atmospheric Physics, Moscow, Russia, 12 November 2020; Volume 11560. [CrossRef]
27. Papkova, A.S.; Shukalo, D.M. CALIPSO stratification of atmospheric aerosols with environmental assessment of dust content over the Black Sea region. *Sovrem. Probl. Distantsionnogo Zondirovaniya Zemli Kosmosa* **2020**, *17*, 234–242. [CrossRef]
28. Barros, R.; Keary, S. Experimental setup for investigation of laser beam propagation along horizontal urban path. *Proc. SPIE* **2014**, *9242*, 396–404.
29. Mata-Calvo, R. Transmitter diversity verification on Artemis geostationary satellite. In Proceedings of the Free-Space Laser Communication and Atmospheric Propagation XXVI, San Francisco, CA, USA, 1–6 February 2014.
30. Mosavi, N.; Marks, B.; Boone, B.; Menyuk, C. Optical beam spreading in the presence of both atmospheric turbulence and quartic aberration. In Proceedings of the Free-Space Laser Communication and Atmospheric Propagation XXVI, San Francisco, CA, USA, 1–6 February 2014.
31. Murty, S.S.R. Laser beam propagation in atmospheric turbulence. *Proc. Indian Acad. Sci.* **1979**, *2*, 179–195. [CrossRef]
32. Kwiecień, J. The effects of atmospheric turbulence on laser beam propagation in a closed space—An analytic and experimental approach. *Opt. Commun.* **2019**, *433*, 200–208. [CrossRef]

33. Searles, S.; Hart, G.; Dowling, J.; Hanley, S. Laser beam propagation in turbulent conditions. *Appl. Opt.* **1991**, *30*, 401–406. [CrossRef]
34. Gareth, D.; Naven, C. Experimental analysis of a laser beam propagating in angular turbulence. *Open Phys.* **2022**, *20*, 402–415.
35. Summerer, L.; Purcell, O. Concepts for wireless energy transmission via laser. *J. Br. Interplanet. Soc.* **2005**, *58*, 1–10.
36. Fahey, T.; Islam, M.; Gardi, A.; Sabatini, R. Laser Beam Atmospheric Propagation Modelling for Aerospace LIDAR Applications. *Atmosphere* **2021**, *12*, 918. [CrossRef]
37. Zohuri, B. Atmospheric Propagation of High-Energy Laser Beams. In *Directed Energy Weapons*; Springer: Cham, Switzerland, 2016.
38. Shuto, Y. Effect of Water and Aerosols Absorption on Laser Beam Propagation in Moist Atmosphere at Eye-Safe Wavelength of 1.57 μm. *J. Electr. Electron. Eng.* **2023**, *11*, 15–22.
39. Rosen, L.; Ipser, J. High Energy Laser Beam Scattering by Atmospheric Aerosol Aureoles. In Proceedings of the Nonlinear Optical Beam Manipulation and High Energy Beam Propagation through the Atmosphere, Los Angeles, CA, USA, 5–20 January 1989.
40. Oosterwijk, A.; Heikamp, S.; Manders-Groot, A.; Lex, A.; Eijk, J. Comparison of modelled atmospheric aerosol content and its influence on high-energy laser propagation. In Proceedings of the Laser Communication and Propagation through the Atmosphere and Oceans VIII, San Diego, CA, USA, 11–15 August 2019.
41. Kudryashov, A.; Rukosuev, A.; Samarkin, V.; Galaktionov, I.; Kopylov, E. Fast adaptive optical system for 1.5 km horizontal beam propagation. In Proceedings of the Unconventional and Indirect Imaging, Image Reconstruction, and Wavefront Sensing 2018, San Diego, CA, USA, 19–23 August 2018.
42. Soloviev, A.; Kotov, A.; Perevalov, S.; Esyunin, M.; Starodubtsev, M.; Alexandrov, A.; Galaktionov, I.; Samarkin, V.; Kudryashov, A.; Ginzburg, V.; et al. Adaptive system for wavefront correction of the PEARL laser facility. *Quantum Electron.* **2022**, *50*, 1115–1122. [CrossRef]
43. Samarkin, V.; Alexandrov, A.; Galaktionov, I.; Kudryashov, A.; Nikitin, A.; Rukosuev, A.; Toporovsky, V.; Sheldakova, Y. Large-aperture adaptive optical system for correcting wavefront distortions of a petawatt Ti:sapphire laser beam. *Quantum Electron.* **2022**, *52*, 187–194. [CrossRef]
44. Buck, A.L. Effects of the Atmosphere on Laser Beam Propagation. *Appl. Opt.* **1967**, *6*, 703–708. [CrossRef] [PubMed]
45. Galaktionov, I.; Kudryashov, A.; Sheldakova, J.; Nikitin, A.; Samarkin, V. Laser beam focusing through the atmosphere aerosol. In Proceedings of the Unconventional and Indirect Imaging, Image Reconstruction, and Wavefront Sensing 2017, San Diego, CA, USA, 6–10 August 2017.
46. Lylova, A.; Kudryashov, A.; Sheldakova, J.; Borsoni, G. The real-time atmospheric turbulence modeling and compensation with the use of adaptive optics. In Proceedings of the Optics in Atmospheric Propagation and Adaptive Systems XVIII, Toulouse, France, 21–24 September 2015.
47. Gonglewski, J.; Highland, R.; Dayton, D.; Sandven, S.; Rogers, S.; Browne, S. ADONIS: Daylight Imaging through Atmospheric Turbulence. In Proceedings of the Digital Image Recovery and Synthesis III, Denver, CO, USA, 4–9 August 1996.
48. Marchi, G.; Weiß, R. Evaluations and Progress in the Development of an Adaptive Optics System for Ground Object Observation. In Proceedings of the Optics in Atmospheric Propagation and Adaptive Systems X, Florence, Italy, 17–20 September 2007.
49. Marchi, G.; Scheifling, C. Adaptive optics concepts and systems for multipurpose applications at near horizontal line of sight: Developments and results. In Proceedings of the Optics in Atmospheric Propagation and Adaptive Systems XII, Berlin, Germany, 31 August–3 September 2009.
50. Segel, M.; Gladysz, S.; Stein, K. Optimal modal compensation in gradient-based wavefront sensorless adaptive optics. In Proceedings of the Laser Communication and Propagation through the Atmosphere and Oceans VIII, San Diego, CA, USA, 11–15 August 2019.
51. Segel, M.; Zepp, A.; Anzuola, E.; Gladysz, S.; Stein, K. Optimization of wavefront-sensorless adaptive optics for horizontal laser beam propagation in a realistic turbulence environment. In Proceedings of the Laser Communication and Propagation through the Atmosphere and Oceans VI, San Diego, CA, USA, 6–10 August 2017.
52. Gonglewski, J.; Dayton, D. Least squares blind deconvolution of air to ground imaging. In Proceedings of the Optics in Atmospheric Propagation and Adaptive Systems VIII, Bruges, Belgium, 19–22 September 2005.
53. Mackey, R.; Dainty, C. Adaptive optics correction over a 3km near horizontal path. In Proceedings of the Optics in Atmospheric Propagation and Adaptive Systems XI, Cardiff, UK, 15–18 September 2008.
54. Belousov, V.; Galaktionov, I.; Kudryashov, A.; Nikitin, A.; Otrubyannikova, O.; Rukosuev, A.; Samarkin, V.; Sivertceva, I.; Sheldakova, J. Adaptive optical system for correction of laser beam going through turbulent atmosphere. *Proc. SPIE* **2020**, *11560*, 1156026.
55. Zakharov, V.E.; L'vov, V.S.; Falkovich, G. *Kolmogorov Spectra of Turbulence I. Wave Turbulence*; Springer: Berlin, Germany, 1992.
56. Rampy, R.; Gavel, D.; Dillon, D.; Thomas, S. Production of phase screens for simulation of atmospheric turbulence. *Appl. Opt.* **2012**, *51*, 8769–8778. [CrossRef]
57. Oh, H.-G.; Kang, P.; Lee, J.; Rhee, H.-G.; Ghim, Y.-S.; Lee, J.H. Fabrication of Phase Plate to Simulate Turbulence Effects on an Optical Imaging System in Strong Atmospheric Conditions. *Curr. Opt. Photonics* **2024**, *8*, 259–269. [CrossRef]

58. Baumer. Passion for Sensors. Available online: https://www.baumer.com/int/en/product-overview/industrial-cameras-image processing/industrial-cameras/c/331 (accessed on 16 September 2024).
59. The Imaging Source. Technologies based on standards. Available online: https://www.theimagingsource.com/en-us/product industrial/23u/dmk23ux174 (accessed on 16 September 2024).
60. Skvortsov, A.A.; Koryachko, M.V.; Zuev, S.M.; Demchenkova, A.A. Melt drops movement over semiconductor surfaces controlle by electric field. *Periodico Tche Quimica* **2019**, *16*, 681–687. [CrossRef]
61. Galaktionov, I.; Nikitin, A.; Samarkin, V.; Sheldakova, J.; Kudryashov, A. Laser beam focusing through the scattering medium-lov order aberration correction approach. In Proceedings of the Unconventional and Indirect Imaging, Image Reconstruction, an Wavefront Sensing 2018, San Diego, CA, USA, 19–23 August 2018.
62. Galaktionov, I.; Kudryashov, A.; Sheldakova, J.; Nikitin, A. Laser beam focusing through the dense multiple scattering suspensio using bimorph mirror. In Proceedings of the SPIE Photonics West 2019, San Francisco, CA, USA, 2–7 February 2019; Volum 10886, p. 1088619.
63. Skvortsov, A.A.; Pshonkin, D.E.; Koryachko, M.V.; Rybakova, M.R. The Effect of Constant Magnetic Fields on the Dynamics o Molten Zones in the Field of Structural Inhomogeneity of Silicon. *Tech. Phys. Lett.* **2018**, *44*, 498–501. [CrossRef]
64. Skvortsov, A.A.; Zuev, S.M.; Koryachko, M.V.; Skvortsova, A.A. The effect of thin dielectric layers at silicon on interconnectio heating dynamics at Thermal Shocks. *Periodico Tche Quimica* **2019**, *16*, 448–456. [CrossRef]
65. Koryachko, M.V.; Pshonkin, D.E.; Skvortsov, A.A. Features of melt droplet formation during electrical destruction aluminum films on the semiconductor surface. *Defect and Diffusion Forum* **2021**, *410*, 737–741. [CrossRef]
66. Skvortsov, A.A.; Koryachko, M.V.; Skvortsov, P.A.; Lukyanov, M.N. On the Issue of Crack Formation in a Thin Dielectric Layer or Silicon under Thermal Shock. *J. Mater. Eng. Perform.* **2020**, *29*, 4390–4395. [CrossRef]
67. Skvortsov, A.A.; Zuev, S.M.; Koryachko, M.V. Non-stationary phase transitions in systems metallization of silicon structures. *Rus Microelectron* **2016**, *45*, 215–222. [CrossRef]
68. Higgins, C.W.; Froidevaux, M.; Simeonov, V.; Vercauteren, N.; Barry, C.; Parlange, M.B. The Effect of Scale on the Applicability o Taylor's Frozen Turbulence Hypothesis in the Atmospheric Boundary Layer. *Bound.-Layer Meteorol* **2012**, *143*, 379–391. [CrossRef
69. Long, Y.; Mo, J.; Wei, X.; Wang, C.; Wang, S. Design of a Moving-magnet Electromagnetic Actuator for Fast Steering Mirro through Finite Element Simulation Method. *J. Magn.* **2014**, *19*, 51–56. [CrossRef]
70. Guignabert, A.; Maillard, T.; Barillot, F.; Sosnicki, O.; Claeyssen, F. Point Ahead Mechanism for Deep Space Optical Communica tion Development of a New Piezo-based Fine Steering Mirror. In Proceedings of the 45th Aerospace Mechanisms Symposium Houston, TX, USA, 13–15 May 2020; pp. 157–161.
71. Csencsics, E.; Sitz, B.; Schitter, G. A Fast Piezo Actuated Tip/Tilt Mirror for raster scan applications. *IFAC-PapersOnLine* **2019**, *52* 289–294. [CrossRef]
72. Skvortsov, A.A.; Zuev, S.M.; Koryachko, M.V.; Voloshinov, E.B. Specific features of motion of molten zones in the field of silicor structural inhomogeneity. *Tech. Phys. Lett.* **2017**, *43*, 705–707. [CrossRef]
73. Platt, B.; Shack, R.J. History and principles of Shack-Hartmann wavefront sensing. *Refr. Surg.* **2001**, *17*, 15. [CrossRef]
74. Liang, J.; Grimm, B.; Goelz, S.; Bille, J.F. Objective measurement of the wave aberrations of the human eye with the use of a Hartmann-Shack wave-front sensor. *J. Opt. Soc. Am. A* **1994**, *11*, 1949. [CrossRef]
75. Lane, R.G. Wave-front reconstruction using a Shack–Hartmann sensor. *Appl. Opt.* **1992**, *31*, 6902. [CrossRef] [PubMed]
76. Primot, J. Theoretical description of Shack–Hartmann wave-front sensor. *Opt. Commun.* **2003**, *222*, 81–92. [CrossRef]
77. Konnik, M.; Doná, J. Waffle mode mitigation in adaptive optics systems: A constrained Receding Horizon Control approach. In Proceedings of the American Control Conference, Washington, DC, USA, 17–19 June 2013; pp. 3390–3396.
78. Mauch, S.; Reger, J. Real-Time Spot Detection and Ordering for a Shack–Hartmann Wavefront Sensor with a Low-Cost FPGA *Proc. IEEE Trans. Instrum. Meas.* **2014**, *63*, 2379–2386. [CrossRef]
79. Galaktionov, I.; Sheldakova, J.; Nikitin, A.; Toporovsky, V.; Kudryashov, A. A Hybrid Model for Analysis of Laser Beam Distortions Using Monte Carlo and Shack–Hartmann Techniques: Numerical Study and Experimental Results. *Algorithms* **2023**, *16*, 337 [CrossRef]
80. Galaktionov, I.; Kudryashov, A.; Sheldakova, J.; Byalko, A.; Borsoni, G. Laser beam propagation and wavefront correction in turbid media. In Proceedings of the Unconventional Imaging and Wavefront Sensing 2015, San Diego, CA, USA, 9–13 August 2015.
81. Galaktionov, I.; Nikitin, A.; Sheldakova, J.; Kudryashov, A. B-spline approximation of a wavefront measured by Shack-Hartmann sensor. In Proceedings of the Laser Beam Shaping XXI, San Diego, CA, USA, 1–5 August 2021.
82. Malacara-Hernandez, D. Wavefront fitting with discrete orthogonal polynomials in a unit radius circle. *Opt. Eng.* **1990**, *29*, 672–675. [CrossRef]
83. Wyant, J.C.; Creath, K. Basic wavefront aberration theory for optical metrology. In *Applied Optics and Optical Engineering*; Academic Press: Cambridge, MA, USA, 1992; pp. 28–39.
84. Genberg, V.; Michels, G.; Doyle, K. Orthogonality of Zernike polynomials. *Proc. SPIE* **2002**, *4771*, 276–287.
85. Lakshminarayanan, V.; Fleck, A. Zernike polynomials: A guide. *J. Mod. Opt.* **2011**, *58*, 545–561. [CrossRef]

86. Linnik, J.V. *Least Square Error Method and the Basics of Mathematical-Statistical Theory*, 2nd ed.; Mir: Moscow, Russia, 1962.
87. Kandidov, V.; Shlenov, A. *Atmospheric Turbulence Simulation. Grid Representation. Technical Summary*; International Laser Center: Moscow, Russia, 1995.
88. Vorontsov, M.; Shmalhausen, V. *Principles of Adaptive Optics*; Nauka: Moscow, Russia, 1985.
89. Dudorov, V.; Kolosov, V.; Filimonov, G. Algorithm for formation of an infinite random turbulent screen. In Proceedings of the Twelfth Joint International Symposium on Atmospheric and Ocean Optics/Atmospheric Physics, Tomsk, Russia, 27–30 June 2005.
90. Lane, R.G.; Glindemann, A.; Dainty, J.C. Simulation of a Kolmogorov phase screen. *Waves Random Media* **1992**, *2*, 209–224. [CrossRef]
91. Galaktionov, I.; Sheldakova, J.; Toporovsky, V.; Kudryashov, A. Atmospheric turbulence with Kolmogorov spectra: Software simulation and real-time reconstruction and compensation by means of adaptive optical system with bimorph and stacked-actuator mirrors. *Photonics* **2023**, *10*, 1147. [CrossRef]
92. Galaktionov, I.; Sheldakova, J.; Kudryashov, A. Wavefront analysis of the laser beam propagating through a turbid medium. *Quantum Electron.* **2015**, *45*, 143–144. [CrossRef]
93. Li, J.; Li, W.; Huang, R. An efficient method for solving a matrix least squares problem over a matrix inequality constraint. *Comput. Opt. Appl.* **2016**, *63*, 393–423. [CrossRef]
94. Galaktionov, I.; Kudryashov, A.; Sheldakova, J.; Byalko, A.; Borsoni, G. Measurement and correction of the wavefront of the laser light in a turbid medium. *Quantum Electron.* **2017**, *47*, 32–37. [CrossRef]
95. Burger, L.; Litvin, I.; Forbes, A. Simulating atmospheric turbulence using a phase-only spatial light modulator. *S. Afr. J. Sci.* **2008**, *104*, 129–137.
96. Laslandes, M.; Patterson, K.; Pellegrino, S. Optimized actuators for ultrathin deformable primary mirrors. *Appl. Opt.* **2015**, *54*, 1559. [CrossRef]
97. Liu, D.; Wang, Z.; Liu, J.; Tan, J.; Yu, L.; Mei, H.; Zhou, Y.; Zhu, N. Performance analysis of 1-km free-space optical communication system over real atmospheric turbulence channels. *Opt. Eng.* **2017**, *56*, 106111. [CrossRef]
98. Li, M.; Cvijetic, M. Coherent free space optics communications over the maritime atmosphere with use of adaptive optics for beam wavefront correction. *Appl. Opt.* **2015**, *54*, 1453–1462. [CrossRef]
99. Wright, M.W.; Morris, J.F.; Kovalik, J.M.; Andrews, K.S.; Abrahamson, M.J.; Biswas, A. Adaptive optics correction into single mode fiber for a low Earth orbiting space to ground optical communication link using the OPALS downlink. *Opt. Express* **2015**, *23*, 33705–33712. [CrossRef]
100. Thompson, C.A.; Kartz, M.W.; Flath, L.M.; Wilks, S.C.; Young, R.A.; Johnson, G.W.; Ruggiero, A.J. Free space optical communications utilizing MEMS adaptive optics correction. In Proceedings of the Free-Space Laser Communication and Laser Imaging II, Seattle, WA, USA, 7–11 July 2002.
101. Cornelissen, S.; Hartzell, A.; Stewart, J.; Bifano, T.; Bierden, T. MEMS Deformable Mirrors for Astronomical Adaptive Optics. In Proceedings of the Adaptive Optics Systems II, San Diego, CA, USA, 27 June–2 July 201.
102. Majumdar, A.K.; Eaton, F.D.; Jensen, M.L.; Kyrazis, D.T.; Schumm, B.; Dierking, M.P.; Shoemake, M.A.; Dexheimer, D.; Ricklin, J.C. Atmospheric Turbulence Measurements over Desert site using groundbased instruments, kite/tethered-blimp platform and aircraft relevant to optical communications and imaging systems: Preliminary Results. In Proceedings of the Free-Space Laser Communications VI, San Diego, CA, USA, 13–17 August 2006.
103. Brady, A.; Berlich, R.; Leonhard, N.; Kopf, T.; Böttner, P.; Eberhardt, R.; Reinlein, C. Experimental validation of phase-only pre-compensation over 494 m free-space propagation. *Opt. Lett.* **2017**, *42*, 2679–2682. [CrossRef]

photonics MDPI

Article

Quantum Dot Waveguide Array for Broadband Light Sources

Dongyang Li [1], Yufei Chu [2,*], Qingbo Xu [1], Dong Liu [1], Junying Ruan [1], Hao Sun [1], Jianwei Li [1], Chengde Guo [1], Xiaoyun Pu [1] and Yuanxian Zhang [1,*]

[1] School of Physics and Astronomy, Yunnan University, Kunming 650500, China; dongyangli2022@itc.ynu.edu.cn (D.L.); xuqingbo@itc.ynu.edu.cn (Q.X.); liudong@stu.ynu.edu.cn (D.L.); tuananh_nguyen@163.com (J.R.); sunhao1@stu.ynu.edu.cn (H.S.); lijianwei_bzfq@stu.ynu.edu.cn (J.L.); guochengde@itc.ynu.edu.cn (C.G.); xypu@163.com (X.P.)

[2] School of Marine and Atmospheric Sciences, Stony Brook University, Stony Brook, NY 11790, USA

* Correspondence: yufei.chu@stonybrook.edu (Y.C.); zyx74635@163.com (Y.Z.)

Abstract: In this paper, we demonstrate a broadband and simultaneous waveguide array light source based on water-soluble CdSe/ZnS quantum dots (QDs). We initially measure the fluorescence intensity for various cladding solution concentrations along the fiber axis to assess their impact on the propagation loss; the experimental results show that the fluorescent intensity decreases with fiber length, with higher concentrations showing a more pronounced decrease. Then, we showcase a synchronous QD light source in an optofluidic chip that fluoresces in red, green, and blue (RGB) within a microfluidic channel. Finally, a 3 × 3 QD array of a fluorescent display on a single PDMS chip is demonstrated. The QD waveguide represents a compact and stable structure that is readily manufacturable, making it an ideal light source for advancing high-throughput biochemical sensing and on-chip spectroscopic analysis.

Keywords: quantum dots; broadband light source; waveguide array; evanescent wave

1. Introduction

An optofluidic system integrates optics and microfluidics in a device to make full use of the advantages of both [1–4]. In optofluidic systems, the optical or optoelectronic properties of the system are adjusted by manipulating the fluid on the micrometer scale, and they also have the advantages of speed, efficiency, and low fluid volume consumption, giving them vast potential for biosensing [5–7], biomedicine [8,9], and chemical analysis [10,11]. As an essential building block to construct an optofluidic system, light sources are one of the main research directions of optofluidics and have attracted intensive interest from various communities. To date, a wide variety of light sources have been integrated into the optofluidic chip and demonstrated as on-chip light sources, including liquid-core/liquid-cladding (L2) waveguides [12–14], liquid-core/air-cladding (LA) waveguides [15], solid-core/liquid-cladding (SL) waveguides [16,17], optofluidic ring resonator (OFRR) laser [18,19], distributed feedback gratings (DFBs) optofluidic laser [20], and Fabry–Perot (F–P) cavities optofluidic laser [21,22]. To achieve an integrated broadband fluorescent light source on a chip, Whitesides et al. generated a combined broadband fluorescent light source using a cascade of single-dye light sources and a parallel array of single-dye light sources [23]. Zhang et al. demonstrated a broadband and simultaneous fluorescent light source under a single excitation light by employing three cascade SL waveguides in conjunction with fluorescence resonance energy transfer of a dye pair, which could cover wavelengths in the 520 nm (green emission) to 710 nm (near-infrared emission) region [16].

Significant advances have been demonstrated in the fields of on-chip light sources and sensors, such as the integration of on-chip photonic devices based on the optical waveguides and optofluidic laser, which was achieved using different organic dyes. Unfortunately, several drawbacks still exist in optofluidic light source devices based on fluorescent dyes, including the following: (1) A majority of optofluidic light source devices are excited in free space, typically in the direction that is perpendicular to the device plane. This pumping scheme suffers from increasing pump background noise, and the spatial distribution of the pump light in the device is nonuniform. Therefore, the fluorescence emission is also nonuniform. (2) Organic fluorescent dye has been the commonly used fluorophore solution for the devices described above. However, organic fluorescent dyes suffer severely from photobleaching, making it very challenging to perform repetitive measurements under the continuous action of a pump light for a long time. Additionally, organic fluorescent dyes are sensitive to the environmental factors of the solvent (pH, viscosity, polarity, and ion concentration), resulting in these waveguides being less attractive in biosensing and chemical detection. In contrast, semiconductor quantum dots (QDs) possess unique advantages over organic dyes [24,25]. Firstly, QDs possess high quantum yields and high resistance to photobleaching in the case of water and other harsh solvents (large pH, strongly acidic, and high ion concentration). Secondly, QDs have a high absorption capability in the blue and UV regions and strong and tunable fluorescence emission properties, which are helpful for achieving different wavelengths of fluorescence (broadband fluorescence light source) by using a single excitation source.

In this work, we use a solid-core/liquid-cladding optofluidic waveguide (SLOFWG) as the platform to demonstrate a broadband and simultaneous fluorescence array light source, which uses water-soluble QDs as the cladding solution. Some advantages possessed by the SLOFWG include are as follows: (i) QDs are used as the cladding solution, which can circumvent various issues toward the organic dyes. (ii) The fluorescence emission in the SLOFWG is generated by the evanescent field of the excitation light. Thanks to the evanescent field of the excitation beam being uniform around the fiber and providing excitation to the QDs, the fluorescence emission from the rim of the fiber is uniform. In addition, the pump light propagates along the fiber due to the TIR at the fiber/cladding interface, which avoids damaging the PDMS substrate. Therefore, this pumping scheme not only effectively prevents the background noise caused by the pump light directly acting on the waveguide materials but also provides a uniform pump along the fiber axis. (iii) The smooth surface of the optical fiber reduces the propagation loss along the fiber axis, which helps to increase the optical path length of the fluorescence emission along the fiber axis (in the microfluidic channel). Therefore, based on this feature, we achieve a simultaneous multi-wavelength fluorescence emission with a single wavelength of the pump via filling different QD solutions into the different microchannels along the axis of the optical fiber. The multi-channel fluorescence emission can be readily scaled up to a fluorescence array to achieve broadband light sources. (iv) The SLOFWG is closed in a PDMS substrate, which can avoid the potential risk to the human body and environment caused by the volatilization of the sample solution, particularly, when using toxic or volatile samples. In the present work, we first investigate the stability of the QDs. In addition, we study in detail the fluorescence intensity of QD solutions with different concentrations along the fiber axis to evaluate the performance of the SLOFWG. The experimental results show that very low propagation loss of the fluorescence intensity can be obtained with the low-concentration QD solution. Second, based on the feature of long-distance fluorescence emission of the SLOFWG, a simultaneous red–green–blue (RGB) fluorescent emission with a single-wavelength excitation light is demonstrated on a PDMS substrate. Finally, the 3 × 3 fluorescence array is constructed and achieved by integrating the SLOFWG in parallel.

2. Materials and Methods

The optical fiber (transmittance per meter > 99%) was provided by Nanjing Chunhui Science and Technology Industrial Co., Ltd. (Nanjing, China). The polydimethylsiloxane (PDMS, RI = 1.405) microfluidic channel was purchased from HICOMP Microtech (Suzhou) Co., Ltd. (Suzhou, China). The CdSe/ZnS QDs were purchased from Suzhou Xingshuo Nanotech Co., Ltd. (Suzhou, China). Disodium fluorescein (DSF) (molecular weight 356.31 g/mol, CAS Number 518-47-8) was obtained from Sigma-Aldrich (St. Louis, Mo, USA). Deionized (DI) water was obtained using an ultrapure system (EU-LS-100TJ, China). All the reagents were used as received without further purification.

Experimental section: The preparation of the PDMS chip is described elsewhere [26]. The size of the chip and the microfluidic channel was 20 mm × 10 mm × 5 mm ($L \times w \times h$) and 15 mm × 0.3 mm × 0.3 mm, respectively. The optical fiber was suspended in the microfluidic channel and was evenly surrounded by the QD aqueous solution. Because the fluorescence emission along the fiber axis (Z-axis) starts from Z = 0 mm, the optical path length (Z) represents the properties (intensity, wavelength, and spectra) of the fluorescence emission along the fiber axis at the Z (≥0 mm) point. The details of the experimental setup have been described earlier in Ref. [26]. Briefly, the liquid inlet or outlet is connected to a micro-pump (Longer Pump, BT100-2J) for fluidic delivery. The QD solution is dispensed into the microchannel and used for the liquid cladding. The QD aqueous solution flowed in the microfluidic channel at a speed of 1 μL/min. A continuous-wavelength (CW) laser of 405 nm was used in all the experiments. The fluorescence emission was recorded by a multimode fiber and then sent to a spectrometer (ICCD, PI-MAX; Spectrometer, Spectrapro 500i) for spectral analysis. The fluorescence emission properties of different locations along the optical fiber axis were recorded by moving the guiding fiber along the fiber axis in free space.

3. Results

3.1. Optical Stability of QDs

The photostability of the luminescent materials is a crucial attribute of the fluorophore because the changes in the fluorescence intensity lead to intrinsic fluctuations.

We performed an experiment to investigate the photostability of the water-soluble CdSe/ZnS QDs and the fluorescence dye disodium fluorescein (DSF). The chemical structures of CdSe/ZnS QDs and DSF are shown in Figure 1a,b. In the experiment, the same concentration (5.0 μM) of QDs and DSF aqueous solution was used as that of the cladding solution. For each cladding solution, we recorded the fluorescence spectra of CdSe/ZnS QDs and DSF under a continuous-wave (CW) laser at 405 nm with an excitation intensity of 5 mW. We then calculated the value of the acquired spectral intensities as a function of the times to obtain the fluorescence intensity traces, as shown in Figure 1c. After continuous and repeated measurements for 600 s, the fluorescence intensity of the CdSe/ZnS QDs experienced a decrease of less than 5%, while a more than 20% decrease was observed for the fluorescence intensity of DSF mainly due to a much stronger photobleaching effect. In addition, we investigated the photostability of the cladding solution under different excitation intensities (10 mW and 20 mW); the photostability of the fluorescence showed very good agreement when the pump power was 5 mW. Therefore, the SLOFWG possesses higher fluorescence stability when the QDs are used as a cladding solution, making it suitable for performing repetitive measurements with a long exposure time.

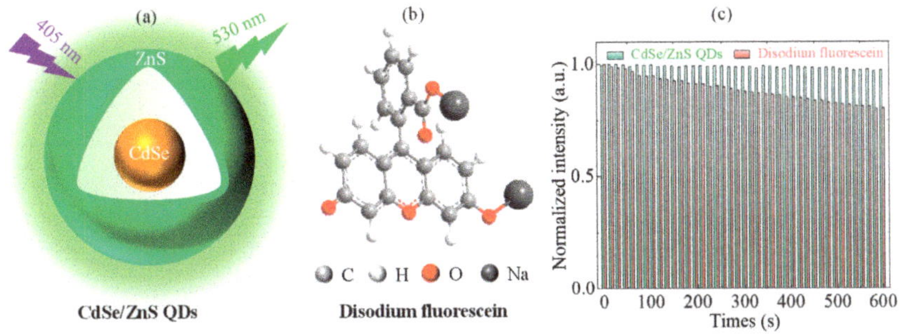

Figure 1. (**a**) Schematic structures of CdSe/ZnS QDs. (**b**) Chemical structures of organic dye, disodium fluorescein. (**c**) Normalized fluorescence intensity of the same concentration of QDs and disodium fluorescein varied with time. The pump power was fixed at 5 mw.

3.2. Light Source of QDs Excited by Evanescent Field

3.2.1. Fluorescence Distribution of SLOFWG at Different Optical Path Lengths

To demonstrate the SLOFWG and realize the integration of the optofluidic light sources, the properties of propagation loss affected by the concentration of the QD solution (cladding solution) were studied. We initially employed a CdSe/ZnS QD aqueous solution with an emission wavelength of 530 nm as the cladding solution. We assumed the fluorescence starting points at Z = 0 mm for each concentration of QDs. The fluorescence spectra of different optical path lengths (Z) along the optical fiber axis were recorded to assess the propagation loss properties of the SLOFWG, as shown in Figure 2a. We also measured the fluorescence spectra for QD concentrations of 1 and 10 μM, respectively. During the measurement, the excitation power was maintained at 5 mW. For the same location, when the concentration of the QD solution increased, the fluorescence emission intensity gradually increased. The corresponding fluorescence spectra intensity is depicted in Figure 2b. Figure 2c presents the calculated values of the fluorescence spectral intensity (in Figure 2b) attenuation as a function of the optical path length using Equation (1). The calculated propagation losses were 0.02, 0.08, and 0.22 dB/cm for three representative QD concentrations of 1, 5, and 10 μM, respectively. The propagation loss of the SLOFWG increased when the concentration of the cladding solution increased. Therefore, we can obtain a low-loss fluorescence emission by using a cladding solution with the appropriate concentration, to achieve uniform fluorescent emission in all the microfluidic channels.

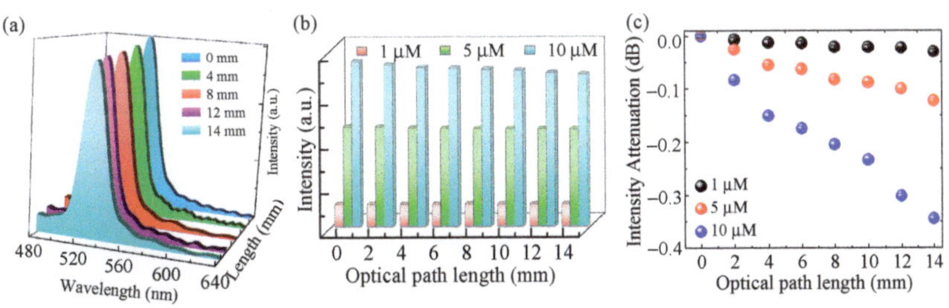

Figure 2. (**a**) Fluorescence spectra of QDs with different path lengths (QD concentration: 5 μM). (**b**) Fluorescence spectra intensity of QDs versus the optical path length (QD concentration of 1, 5, 10 μM, respectively). (**c**) Fluorescence intensity attenuation with various optical path lengths. The error bars are calculated with five repeated measurements.

The fluorescence spectral intensity attenuation (A) of the SLOFWG along the Z-axis can be expressed by the following formula [27]:

$$A = 10 \cdot \log_{10}\left(\frac{I_z}{I_{z=0}}\right) \quad (1)$$

where I_Z represents the fluorescence spectral intensities of QDs at Z $\geq$ 0 mm, and $I_{Z=0}$ is the fluorescence spectral intensities of QDs at Z = 0 mm. $I_{Z=0} = CN\lambda p/[2\pi(n_{2_1} \sin^2\theta_t - n_{2_2})^{1/2} + \lambda_p \alpha^{P}_{abs,out}]$ [28]. Therefore, Equation (1) can be expressed as follows:

$$A = -10\left[\eta_P \alpha^{P}_{abs,out} + \alpha^{P}_{abs,in}(1-\eta_P)\right] \log_{10} e \cdot Z \quad (2)$$

where $\alpha^{P}_{abs,in}$ and $\alpha^{P}_{abs,out}$ represent the coefficient of fiber absorption to the excitation light and the cladding solution absorption coefficient to the excitation light, respectively, and η_P is the modal volume ratio.

The absorption coefficient ($\alpha^{P}_{abs,in}$) of the optical fiber was 0.0002 cm^{-1} in our experiment [28]. Moreover, $\alpha^{P}_{abs,out} = N\sigma_{abs}$, where N is the concentration of the QDs, σ_{abs} is the absorption cross section (σ_{abs}) of the CdSe/ZnS QDs. The σ_{abs} of CdSe/ZnS QDs at λ_P = 405 nm was 4.00 × 10^{-14} cm^2. Therefore, $\alpha^{P}_{abs,out}$ increased when the concentration of the QDs was in the range of 1.0 to 10 μM. Because direct measurement of the η_P is challenging, η_P values in the range of 0.0001 to 0.0005 were scanned in the simulation, and the results on the propagation losses dependent on optical path length were compared to the experimental results to determine the η_P of our SLOFWG.

When the concentration of the cladding solution was in the range of 1 μM to 10 μM and the RI was fixed at 1.333, the fitting η_P value remained unchanged. However, when the concentration of the QDs increases, more QDs are excited by the pump light. Thus, the loss of absorption and the fluorescence emission intensity are increased simultaneously. In addition, the increase in the absorption coefficient (α^{P}_{abs}) will result in an increase in the propagation loss. Consequently, a higher concentration of the cladding solution will lead to a large propagation loss, as shown in Figure 2c.

3.2.2. Simultaneous Broadband Light Sources

A simultaneous and full-color fluorescent light source with a single pump light provides an ideal light source for high-throughput biochemical sensing and on-chip spectroscopic analysis.

As described above, due to the properties of low propagation loss of the SLOFWG along the fiber axis, it provides an excellent tool for achieving simultaneous and broadband light sources. However, when different organic dyes are directly excited by a single pump light, the limited choice of dyes can either obtain just a narrow range of emission wavelength range or cover the desired broad spectral output range with low pump efficiency. Although the low pump efficiency may be alleviated by using a higher dye concentration, the existence of dye self-absorption severely compromises the fluorescence performance. QDs overcome the shortcomings of the organic dyes, due to their high light absorption capability in the blue and UV regions. To achieve a high pump efficiency and broaden the range of the spectral output, QD aqueous solutions with different emission wavelengths are used as cladding solutions.

In this section, we demonstrate a simultaneous and multi-wavelength fluorescence light source in the case of a single pump light, which is used as the feature of a long-distance fluorescence emission based on evanescent wave excitation. To this end, we designed a PDMS-based microfluidic chip with three channels, as shown in Figure 3a. The chip consisted of three cascaded SLOFWGs of the same size

($L \times w \times h$ = 2 mm × 0.3 mm × 0.3 mm). The chip contained three inlet ports (injecting the cladding solutions) and three outlet ports (outflowing the cladding solution), respectively. The size of the chip port was 125 μm, which is the same as the fiber diameter, ensuring the three SLOFWGs are parallel in alignment. To eliminate the diffusion between the cladding solution of each SLOFWG, the spacing between the two adjacent fluidic channels was 8 mm. Figure 3b displays the absorption of CdSe/ZnS QDs with different emission wavelengths; all the CdSe/ZnS QDs exhibited strong absorption at 405 nm (purple dotted line in Figure 3b).

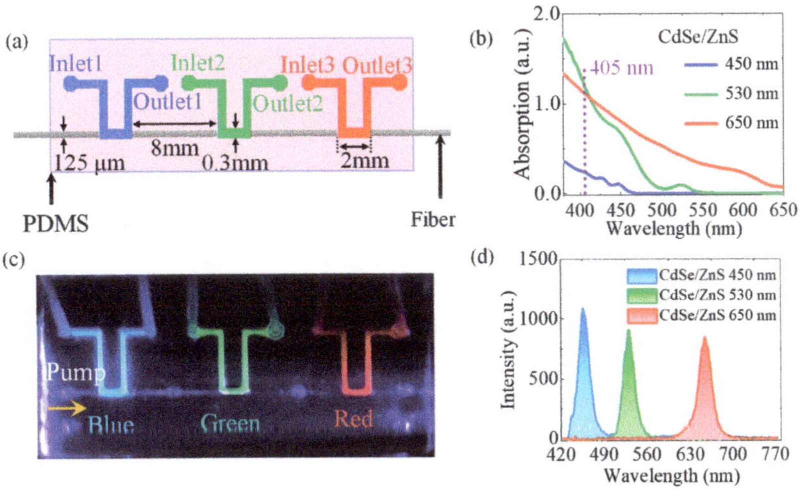

Figure 3. (**a**) The schematic design diagram of the PDMS optofluidic device for a red–green–blue three-color fluorescent light source. (**b**) The absorption of CdSe/ZnS QDs with different emission wavelengths. (**c**) Image of the red–green–blue three-color fluorescent light source in operation. (**d**) The spectra of fluorescence emission of the broadband light source from different channels. The blue, green, and red shades are the fluorescence spectra of CdSe/ZnS QDs (emission wavelengths of 450 nm, 530 nm, and 650 nm) with a concentration of 5 μM, respectively.

In the experiment, the aqueous solutions of the CdSe/ZnS QDs (emission wavelengths 450 nm, 530 nm, and 650 nm) were injected into channels 1, 2, and 3 using micro-pumps, respectively. The flow rate was set at a speed of 1 μL/min to allow the QD aqueous solution to go through the microfluidic channel. A CW laser with the wavelength 405 nm was used as the pump light and longitudinally coupled into the fiber along the fiber axis; three different wavelengths and nearly uniform fluorescent emission were achieved in the same PDMS substrate under a single excitation wavelength. As shown in Figure 3c, a clear image of the three red–green–blue fluorescent light sources can be observed simultaneously. The spectra of three of the red–green–blue fluorescent lights were also picked up, as shown in Figure 3d.

3.3. A 3 × 3 Optofluidic Fluorescence Array

Optofluidic fluorescence arrays hold great promise for on-chip broadband light sources, full-color display, and high-throughput biochemical sensing. To further characterize the properties of our SLOFWG, we demonstrated the simultaneous 3 × 3 fluorescence array under a single pump light. The array chip was constructed by integrating an SLOFWG with the same size in parallel in an arrayed PDMS chip. To prevent diffusion between the QD solution among each SLOFWG, there was a spacing between the two adjacent fluidic channels, which was set to 9 mm. The chip contained nine cascaded and parallel SLOFWGs with the same size ($l \times w \times h$ = 1 mm × 1 mm × 500 μm). The size of the fiber port was 125 μm, which was the same as the fiber diameter, ensuring the fluorescence array was parallel in alignment, as shown in Figure 4a. The optofluidic array chip image is shown in Figure 4b.

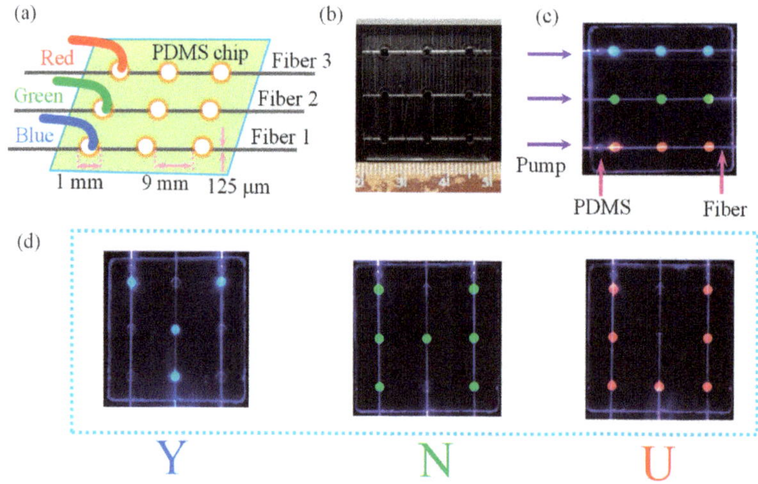

Figure 4. (**a**) The design diagram for a 3 × 3 fluorescence array. The red, green, and blue pipes represent the 450 nm, 530 nm, and 650 nm CdSe/ZnS QD solution injections, respectively. (**b**) Photograph of PDMS chip with 3 × 3 fluorescence array. (**c**) Photograph of RGB fluorescence emission by using a 3 × 3 fluorescence array. (**d**) Photograph of a fluorescence display prototype showing "YNU" based on a 3 × 3 fluorescence array.

For the experiment, three distinct CdSe/ZnS QD aqueous solutions with different emission wavelengths, 450 nm, 530 nm, and 650 nm, were selected to demonstrate the full-color display using the 3 × 3 fluorescence array chip. First, the three CdSe/ZnS QD aqueous solutions were injected into the channel using a micro-pump. A laser diode (wavelength 405 nm) was used as the pump source, and the pump power was fixed at 10 mW. The pump light was converged into a linear spot through a cylindrical lens (0.5 mm × 6 cm) and longitudinally coupled into the fiber along the fiber axis to excite the 3 × 3 fluorescence array. When each row channel was injected with CdSe/ZnS QD aqueous solutions with the same emission wavelength, the red–green–blue fluorescence emission with a row can be observed on the array chip, as shown in Figure 4c. In addition, by injecting quantum dot solutions with different emission wavelengths into different channels, different colors of fluorescence can be developed at different channel positions, to obtain images with different characters, and the 3 × 3 array is used to realize the color rendering of "YNU", as shown in Figure 4d.

In the present study, we created an on-chip waveguide array of broadband light sources using the feature of long-distance fluorescence emission based on evanescent wave excitation. The waveguide array provides a useful platform for on-chip light source and biochemical detection; broadband light sources have been studied in depth [29,30]. Multi-parameter sensing can be achieved based on broadband light sources. The multi-parameter sensing chip can be used for heavy ion detection and biological diagnostics [7]. Considering the richness of fluorescent materials and the different types of new fluorophores being developed, fluorophores with low toxicity and low environmental impact can be used in the waveguide array to achieve a full-color light source.

4. Conclusions

In summary, we demonstrated a broadband and simultaneous fluorescence array light source based on SLOFWG. The approach of evanescent wave excitation along the fiber axis avoids background noise and achieves a uniform fluorescence emission. We first studied the stability of QDs. Then, the effect of the cladding solution concentration on the fluorescence emission intensity was investigated to evaluate the performance of the SLOFWG. The results showed that a higher cladding concentration will result in a large attenuation. In addition, we designed and fabricated a microfluidic chip with three channels on a PDMS substrate and achieved a simultaneous red–green–blue fluorescence emission without changing the excitation wavelength by injecting different QDs in three channels, respectively. Finally, the fluorescence array was simultaneously achieved by integrating three channels on a chip with a size of 3 cm $\times$ 3 cm. The 3 $\times$ 3 simultaneous fluorescent arrays can be extended out to an N $\times$ M fluorescent array (N and M are integers), which provides an ideal light source for simultaneous multi-parameter sensing. The simultaneous multi-parameter sensing chip can be utilized for biochemical analytes.

Author Contributions: Conceptualization, Y.Z.; methodology, Y.Z. and D.L. (Dongyang Li); software, D.L. (Dongyang Li), J.L. and C.G.; experiment, D.L. (Dongyang Li), H.S. and D.L. (Dong Liu); data curation, D.L. (Dongyang Li), J.R. and Q.X.; writing—original draft preparation, D.L. (Dongyang Li) and Y.Z.; writing—review and editing, Y.Z., X.P. and Y.C.; supervision, Y.Z.; funding acquisition, Y.Z. and D.L. (Dongyang Li). All authors have read and agreed to the published version of the manuscript.

Funding: This work was supported by the National Natural Science Foundation of China (12404443), the Natural Science Foundation of Yunnan Province of China (202301BF070001-002), the Xingdian Talents Program of Yunnan Province (YNWR-QNBJ-2019-149), the Caiyun Postdoctoral Project (C615300504057), the China Scholarship Council (202307030044), and the Scientific Research Project of Postgraduate Students of Yunnan University (KC-23235352, KC-23235361).

Institutional Review Board Statement: Not applicable.

Informed Consent Statement: Not applicable.

Data Availability Statement: The data underlying the results presented in this paper are not publicly available at this time but may be obtained from the authors upon reasonable request.

Conflicts of Interest: The authors declare no conflicts of interest.

References

1. Zhang, W.; Zhang, Z.; Li, J.; Jiang, S.; Fang, X.; Li, Z. An integrated optofluidic droplet lens driven by a fast thermoelectric cooler. *Opt. Lasers Eng.* **2023**, *169*, 107689. [CrossRef]
2. Chen, Q.; Li, T.; Zhu, Y.; Yu, W.; Zhang, X. Dielectrophoresis-actuated in-plane optofluidic lens with tunability of focal length from negative to positive. *Opt. Express* **2018**, *26*, 6532–6542. [CrossRef] [PubMed]
3. Psaltis, D.; Quake, S.R.; Yang, C. Developing optofluidic technology through the fusion of microfluidics and optics. *Nature* **2006**, *442*, 381–386. [CrossRef]
4. Zhang, Y.; Watts, B.R.; Guo, T.; Zhang, Z.; Xu, C.; Fang, Q. Optofluidic device based microflow cytometers for particle/cell detection: A review. *Micromachines* **2016**, *6*, 70. [CrossRef] [PubMed]
5. Testa, G.; Persichetti, G.; Bernini, R. Optofluidic biosensing: Devices, strategies, and applications. *Trends Anal. Chem.* **2024**, *178*, 117865. [CrossRef]
6. Liu, Y.; Zhang, Y.; Wang, C.; Wang, Y.; Zhang, K.; Yang, X.; Peng, G.-D.; Liu, S.; Wang, Z.; Rao, Y.-J.; et al. Multiplexed optofluidic laser immunosensor for sensitive and rapid detection of biomarkers. *Sens. Actuator B-Chem.* **2024**, *403*, 135198. [CrossRef]
7. Yang, X.; Luo, Y.; Liu, Y.; Gong, C.; Wang, Y.; Rao, Y.-J.; Peng, G.-D.; Gong, Y. Mass production of thin-walled hollow optical fibers enables disposable optofluidic laser immunosensors. *Lab Chip* **2020**, *20*, 923–930. [CrossRef]
8. Yang, D.; Hu, Q.; Zhao, S.; Hu, X.; Gao, X.; Dai, F.; Zheng, Y.; Yang, Y.; Cheng, Y. An optofluidic system for the concentration gradient screening of oocyte protection drugs. *Talanta* **2024**, *278*, 126472. [CrossRef] [PubMed]

9. Li, Z.; Liang, L.; Lin, W.; Huang, Y.; Huang, T.; Wang, W.; Ma, J.; Li, J.; Sun, L.-P.; Guan, B.-O. Optofluidic laser sensor for the detection of dopamine. *Sens. Actuator B-Chem.* **2023**, *390*, 133941. [CrossRef]
10. Gong, C.; Gong, Y.; Oo, M.; Wu, Y.; Rao, Y.-J.; Tan, X.; Fan, X. Sensitive sulfide ion detection by optofluidic catalytic laser using horseradish peroxidase (HRP) enzyme. *Biosens. Bioelectron.* **2017**, *96*, 351–357. [CrossRef]
11. Li, D.; Xu, Q.; Zhao, F.; Guo, C.; Li, J.; Pu, X.; Xia, J.; Lü, Y.; Zhang, Y. Highly sensitive and selective detection of nitrite using a fiber optofluidic laser. *Opt. Express* **2023**, *31*, 31982–31992. [CrossRef] [PubMed]
12. Lee, K.; Yoon, S.; Lee, K.; Kim, S.; Sung, H.; Kim, S. Optofluidic particle manipulation in a liquid-core/liquid-cladding waveguide. *Opt. Express* **2012**, *20*, 17348–17358. [CrossRef] [PubMed]
13. Wang, X.; Song, L.; Wang, X.; Lu, S.; Zheng, Y.; Fang, F. All-optical high-resolution on-chip pressure sensing method based on tunable liquid-core/liquid-cladding waveguide. *IEEE Sens. J.* **2024**, *24*, 1287–1295. [CrossRef]
14. Fan, S.-K.; Lee, H.-P.; Chien, C.-C.; Lu, Y.-W.; Chiu, Y.; Lin, F.-Y. Reconfigurable liquid-core/liquid-cladding optical waveguides with dielectrophoresis-driven virtual microchannels on an electromicrofluidic platform. *Lab Chip* **2016**, *16*, 847–854. [CrossRef] [PubMed]
15. Lim, J.-M.; Kim, S.-H.; Choi, J.-H.; Yang, S.-M. Fluorescent liquid-core/air-cladding waveguides towards integrated optofluidic light sources. *Lab Chip* **2008**, *8*, 1580–1585. [CrossRef]
16. Zhang, Y.; Kenarangi, F.; Zhang, H.; Vaziri, S.; Li, D.; Pu, X.; Sun, Y. Versatile optofluidic solid-core/liquid-cladding waveguide based on evanescent wave excitation. *Anal. Chem.* **2020**, *92*, 14983–14989. [CrossRef]
17. Chu, Y.; Zhang, Y.; Liu, C.; Pu, X. Fluorescence radiation characteristics based on evanescent wave pumping in a microfluidic chip. *Acta Phys. Sin.* **2017**, *66*, 104208. [CrossRef]
18. Li, D.; Zhou, L.; Yu, Q.; Pu, X.; Sun, Y.; Zhou, Q.; Zhang, Y. Optical fiber optofluidic laser based on surfactant solubilization of rhodamine B gain in an aqueous solution. *Opt. Express* **2022**, *30*, 23295–23304. [CrossRef]
19. Lee, W.; Kim, D.; Song, M.; Yoon, D. Optofluidic ring resonator laser with an edible liquid laser gain medium. *Opt. Express* **2017**, *25*, 14043–14048. [CrossRef] [PubMed]
20. Sano, T.; Black, J.; Mitchell, S.; Zhang, H.; Schmidt, H. Pneumatically tunable optofluidic DFB dye laser using corrugated sidewalls. *Opt. Lett.* **2020**, *45*, 5978–5981. [CrossRef]
21. Qiao, Z.; Gong, C.; Liao, Y.; Wang, C.; Chan, K.; Zhu, S.; Kim, M.; Chen, Y.-C. Tunable optical vortex from a nanogroove-structured optofluidic microlaser. *Nano Lett.* **2022**, *22*, 1425–1432. [CrossRef] [PubMed]
22. Yuan, Z.; Huang, S.-H.; Qiao, Z.; Wu, P.; Chen, Y.-C. Metasurface-tunable lasing polarizations in a microcavity. *Optica* **2023**, *10*, 269–278. [CrossRef]
23. Mayers, B.T.; Vezenov, D.V.; Vullev, V.I.; Whitesides, G.M. Arrays and cascades of fluorescent liquid-liquid waveguides: Broadband light sources for spectroscopy in microchannels. *Anal. Chem.* **2005**, *77*, 1310–1316. [CrossRef] [PubMed]
24. Wang, Y.; Leck, K.S.; Ta, V.D.; Chen, R.; Nalla, V.; Gao, Y.; He, T.; Demir, H.V.; Sun, H. Blue liquid lasers from solution of CdZnS/ZnS ternary alloy quantum dots with quasi-continuous pumping. *Adv. Mater.* **2015**, *27*, 169–175. [CrossRef] [PubMed]
25. Chen, Q.; Kiraz, A.; Fan, X. Optofluidic FRET lasers using aqueous quantum dots as donors. *Lab Chip* **2016**, *16*, 353–359. [CrossRef] [PubMed]
26. Zhang, Y.; Li, D.; Ou, Y.; Pu, X.; Sun, Y. The study on lasing threshold properties of rhodamine B in glycerol aqueous solution. *IEEE Photonics J.* **2019**, *11*, 1501709. [CrossRef]
27. Yu, J.; Xu, J.; Dong, Q.; Qi, J.; Chen, J.; Zhang, A.; Song, Y.; Chen, W.; Cheng, Y. Low-loss optofluidic waveguides in fused silica enabled by spatially shaped femtosecond laser assisted etching combined with carbon dioxide laser irradiation. *Opt. Laser Technol.* **2023**, *158*, 108889. [CrossRef]
28. Zhang, Y.; Pu, X.; Zhu, K.; Feng, L. Threshold property of whispering-gallery-mode fiber lasers pumped by evanescent waves. *J. Opt. Soc. Am. B* **2011**, *28*, 2048–2056. [CrossRef]
29. Wang, C.; Gong, C.; Zhang, Y.; Qiao, Z.; Yuan, Z.; Gong, Y.; Chang, G.; Tu, W.; Chen, Y. Programmable rainbow-colored optofluidic fiber laser encoded with topologically structured chiral droplets. *ACS Nano* **2021**, *15*, 11126–11136. [CrossRef]
30. Zhao, J.; Yan, Y.; Gao, Z.; Du, Y.; Dong, H.; Yao, J.; Zhao, Y. Full-color laser displays based on organic printed microlaser arrays. *Nat. Commun.* **2019**, *10*, 870. [CrossRef] [PubMed]

photonics

MDPI

Communication

Modeling a Fully Polarized Optical Fiber Suitable for Photonic Integrated Circuits or Sensors

Wenbo Sun

Donghai Laboratory, Zhoushan 316000, China; haitianyise@donghailab.com

Abstract: A method is developed to make an optical fiber that only transmits fully linearly polarized light and maintains the polarization state. The method for efficient ingesting laser into this fiber is also reported. Using an optical fiber with a prism head, we can compress a plane wave into the thin rectangular cross-section fiber, and the light intensity within the fiber is much larger than that of the incidence wave. Our finite-difference time-domain (FDTD) simulation results show that the compressed light in the fiber becomes fully polarized and maintains the polarization state, and can be well coupled out by the resonance rings. This method is suitable for developing highly efficient polarization-maintaining optical fibers in a much simpler way, for applications in photonic integrated circuits or optical sensors.

Keywords: fiber optics and optical communications; polarization-maintaining fibers; photonic integrated circuits; optical sensors

1. Introduction

Polarization-maintaining fibers (PMFs) that allow the input light to propagate only in one polarization mode have many applications in optical communication systems [1,2], optical fiber interferometry [3], and optical devices requiring high polarization stability, such as modulators. Optimizing the polarization stability of PMFs [4,5] is one of the critical considerations in developing PMF-dependent systems. Traditionally, the PMFs' characteristic of maintaining the polarization throughout the propagation process is achieved by inducing stresses in the material itself during the manufacturing phase [1]. That is, apart from the fiber core, this fiber also contains stress rods that can cause a strong built-in birefringence so that only one polarization state of light can transmit through the fiber. Thus, when PMFs are terminated using fiber connectors, stress rods must align with the connector key correctly. Also, when it comes to splicing, the alignment of the stress rods needs great care. These make the development of the optical system difficult. The second critical consideration in developing PMF-dependent systems is to improve the efficiency of coupling PMFs to laser beams [6,7], including using beam-shaping optics. Our report will focus on these two issues in this study.

In Section 2, the method to efficiently ingest and polarize a plane light wave into an optical fiber is introduced. The finite-difference time-domain (FDTD) method [8–10] with a uniaxial perfectly matched layer (UPML) absorbing boundary condition (ABC) [11] is used to simulate the method's effect on collecting and transmitting polarized light through a fiber. Numerical results are given in Section 3. And in Section 4, we give a summary and conclusions.

2. Method

To efficiently couple and polarize a plane light wave into an optical fiber, we assumed an optical fiber with a prism receiving head, as shown in Figure 1. The optical fiber had a rectangular cross-section; its width was λ and thickness was $\lambda/2$, where λ denotes the light's wavelength inside the optical material of the fiber. The prism receiving head had the same optical properties and width as the fiber. The length (*l*) and the half aperture angle (θ) of the receiving prism of the fiber determines the coupling efficiency, i.e., the wave amount received by the system. However, the light intensity in the fiber is also affected by the refractive index of the fiber, the aperture angle, and the polarization status of the incidence, due to the backscatter from the oblique surface of the prism, and the polarization characteristic of the fiber. In this study, we assumed a fiber of silicon with incidence light at the near infrared (IR) wavelength of λ_0 = 1.5 μm. Utilizing the refractive index of 3.5 of silicon at this wavelength, the wavelength inside the silicon was $\lambda = \lambda_0/3.5$, and the half aperture angle of the fiber's receiving head (θ) was set to 3.5°. The length (*l*) of the receiving head was set to only 5.5 λ. We used the finite-difference time-domain (FDTD) method [8–10] to simulate light's interaction with this system. The FDTD spatial cubic cell size was set to $\delta = \lambda/10$ and the time step was set to $\delta t = \delta/(2c)$, where *c* denotes light speed in free space.

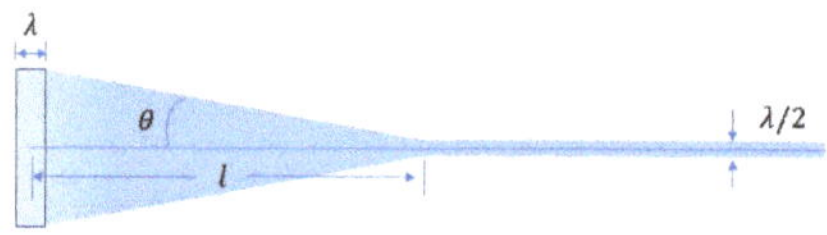

Figure 1. Illustration of the optical fiber and its width, thickness, and the half aperture angle of its receiving end.

The fiber and the receiving head prism are positioned in the FDTD computational domain as illustrated in Figure 2. We set periodic boundary conditions (PBC) [12] at 4 sides of the computational domain around the fiber to simulate a large array of this system. Along the light's incidence direction, the computational domain was truncated by two 6-spatial-cell-thick uniaxial perfectly matched layer (UPML)-absorbing boundaries [11]. The fiber is effectively an infinitely long one since its other end is inside the UPML, which has negligible reflection. A profile of the material property of the computational domain is shown in Figure 3. The perfect conductor is arranged around the receiving head prism to block light leakage before entering the rectangular fiber. The lower fiber has identical geometric parameters and optical properties to those of the upper one. The two resonance rings have the same refractive index as that of the fibers. The upper ring's central radius (i.e., the radius at the ½ thickness of the ring) is *17* λ, and the lower ring's central radius is *13* λ. The radii of the two rings were arbitrarily chosen as exemplary cases. The radii of the rings affect the coupling length of the fibers, thus exchanging the position of the two rings could change the resonance effect. The distance between the outer edge of the ring and the edge of the straight fiber is one FDTD cubic cell δ. All the optical fibers are on the same *z* plane.

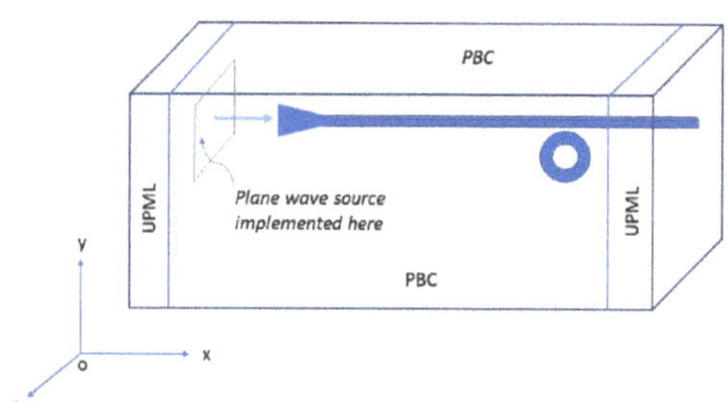

Figure 2. Illustration of the optical fiber, light source implementing plane, uniaxial perfectly matched layer (UPML)-absorbing boundary conditions (ABC) and periodic boundary conditions (PBC) in the computational domain of the FDTD method.

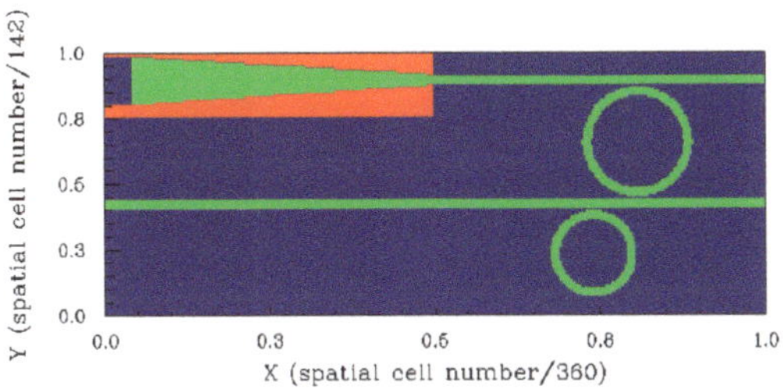

Figure 3. The material refractive index profile in the computational domain. The green color denotes fiber material, the red color denotes perfect conductor, and the blue color denotes free space.

3. Numerical Results

We implemented a plane wave with electric field amplitude of *1* at the wave source plane as illustrated in Figure 2. At first, we simulated a y-polarized light's incidence on the system. Figure 4 shows that, after 300 FDTD time steps, the wave cannot effectively enter the fiber and the light's field inside the fiber is nearly zero. This is because in the y direction the fiber has only a half-wavelength thickness, reaching the limit of diffraction, and the electric field oscillating in the y direction and reflected by the fiber edge has a mostly destructive phase difference inside the fiber, thus light cannot propagate.

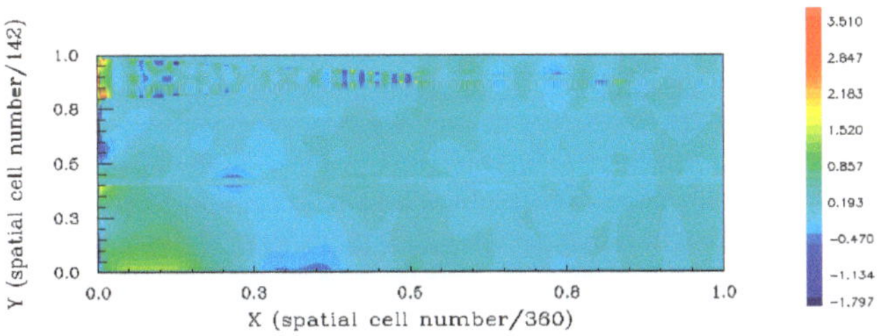

Figure 4. The finite-difference time-domain (FDTD) result for incidence plane light fully polarized in the y direction (E_y only). After 300 time steps of simulation, the light intensity inside the fiber is still nearly zero, which means the y-direction polarized light cannot enter the fiber effectively.

We then simulated a z-polarized light's incidence on the system. Figure 5 shows the numerical results after 600 FDTD time steps. We can see that the wave is well ingested into the rectangular fiber. The amplitude of the electric field can be as large as ~*2.0*. This means the light's intensity inside the rectangular fiber is about four times that of the incidence. The micron-sized receiving head prism works very efficiently to receive light. Additionally, the light entering the fiber keeps good mode and can be well coupled out by resonance rings. It means that this design can be applied in a photonic integrated circuit device. Also, comparing the results in Figures 4 and 5, we can conclude that this system can well polarize the incident light, i.e., only z-polarized light can transfer efficiently in the system. This can be applied to make polarization-maintaining fibers in a much simpler way than that of using stress rods [1]. Note that our simulations of the resonance rings' central radii are simply set as integer numbers of wavelengths inside the macro silicon material (i.e., $\lambda = \lambda_0/3.5$), not the effective wavelength based on effective indices (e.g., [13]) of the specific resonance rings. As such, the light is not completely coupled out and there is still some residue propagation inside the straight fibers after the coupling processes. Note here that in these modeling studies, we assume a plane incidence wave for simplicity; since the fiber is very thin, this should not cause significant difference from the use of a Gaussian beam.

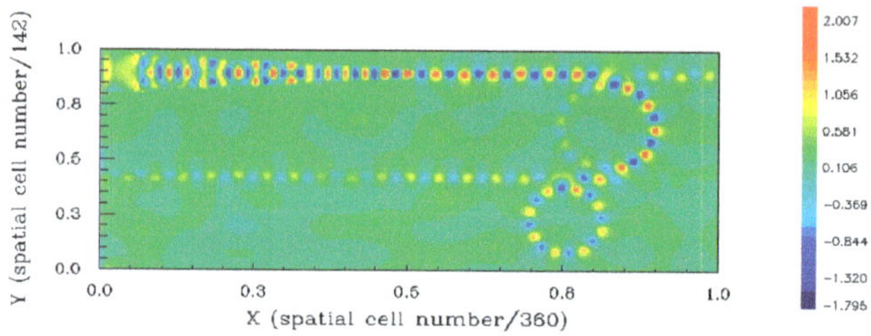

Figure 5. The finite-difference time-domain (FDTD) result for incidence plane light fully polarized in z direction (E_z only). After 600 time steps of simulation, the compressed light in the fiber keeps good transfer mode and can be well coupled out by resonance rings. The light intensity inside the fiber is ~4 times of that of incidence.

4. Conclusions

In this study, a method is developed to compress and polarize a plane light wave into an optical fiber. Using an optical fiber with a prism receiving head, we can compress plane wave into a thin rectangular cross-section fiber, and the light intensity inside the fiber is much larger than the incidence wave. We assume a fiber of silicon with incidence light at the near infrared (IR) wavelength of *1.5 µm*. Utilizing a refractive index of 3.5 of the silicon material at this wavelength, the half aperture angle of the fiber's receiving head (θ) is set to 3.5°. The fiber thickness and width are set as a half wavelength and one wavelength, respectively (the wavelength here is the light wavelength in the fiber material). Our FDTD simulation results show that the compressed light in the fiber becomes well polarized and keeps a good transfer mode that can be well coupled out by resonance rings. The light intensity inside the fiber can be changed by adjusting the length of the prism while preserving the aperture angle. This method is suitable for developing highly efficient polarization-maintaining optical fibers in a much simpler way, for applications in photonic integrated circuits or optical sensors. Note here that this is only a modeling study. But, experiments can be done by depositing a silicon film on a copper substrate [14] and photoetching it following the shape and size of this design.

Funding: This research received no external funding.

Data Availability Statement: Data underlying the results presented in this paper are available on the computer of Donghai Laboratory.

Acknowledgments: The study is supported by the Science Foundation of Donghai Laboratory.

Conflicts of Interest: The author declares no conflict of interest.

References

1. Noda, J.; Okamoto, K.; Sasaki, Y. Polarization-maintaining fibers and their applications. *J. Light. Technol.* **1986**, *4*, 1071–1089. [CrossRef]
2. Xu, F.; Ma, X.; Zhang, Q.; Lo, H.-K.; Pan, J.-W. Secure quantum key distribution with realistic devices. *Rev. Mod. Phys.* **2020**, *92*, 025002. [CrossRef]
3. Pan, J.W.; Chen, Z.B.; Lu, C.Y.; Weinfurter, H.; Zeilinger, A.; Żukowski, M. Multiphoton entanglement and interferometry. *Rev. Mod. Phys.* **2012**, *84*, 777–838. [CrossRef]
4. Walker, G.; Walker, N. Alignment of polarisationmaintaining fibres by temperature modulation. *Electron. Lett.* **1987**, *23*, 689–691. [CrossRef]
5. Arora, P.; Agarwal, A.; Gupta, A.S. Simple alignment technique for polarisation maintaining fibres. *Rev. Sci. Instrum.* **2011**, *82*, 125103. [CrossRef] [PubMed]
6. Nyman, R.A.; Varoquaux, G.; Villier, B.; Sacchet, D.; Moron, F.; Le Coq, Y.; Aspect, A.; Bouyer, P. Tapered-amplified antireflectioncoated laser diodes for potassium and rubidium atomic-physics experiments. *Rev. Sci. Instrum.* **2006**, *77*, 033105. [CrossRef]
7. Bolpasi, V.; Von Klitzing, W. Double-pass tapered amplifier diode laser with an output power of 1 W for an injection power of only 200 µW. *Rev. Sci. Instrum.* **2010**, *81*, 113108. [CrossRef] [PubMed]

. Yee, K.S. Numerical solution of initial boundary value problems involving Maxwell's equations in isotropic media. *IEEE Trans. Antennas Propag.* **1966**, *14*, 302–307.

. Sun, W.; Fu, Q.; Chen, Z. Finite-difference time-domain solution of light scattering by dielectric particles with a perfectly matched layer absorbing boundary condition. *Appl. Opt.* **1999**, *38*, 3141–3151. [CrossRef] [PubMed]

0. Sun, W.; Loeb, N.G.; Fu, Q. Finite-difference time-domain solution of light scattering and absorption by particles in an absorbing medium. *Appl. Opt.* **2002**, *41*, 5728–5743. [CrossRef] [PubMed]

1. Sacks, Z.; Kingsland, D.; Lee, R.; Lee, J.-F. A perfectly matched anisotropic absorber for use as an absorbing boundary condition. *IEEE Trans. Antennas Propag.* **1995**, *43*, 1460–1463. [CrossRef]

2. Sun, W.; Videen, G.; Lin, B.; Hu, Y. Modeling light scattered from and transmitted through dielectric periodic structures on a substrate. *Appl. Opt.* **2007**, *46*, 1150–1156. [CrossRef] [PubMed]

3. Savolainen, J.-M.; Grüner-Nielsen, L.; Kristensen, P.; Balling, P. Measurement of effective refractive-index differences in a few-mode fiber by axial fiber stretching. *Opt. Express* **2012**, *20*, 18646–18651. [CrossRef] [PubMed]

4. Jasinski, J.M.; Gates, S.M. Silicon chemical vapor deposition one step at a time: Fundamental studies of silicon hydride chemistry. *Acc. Chem. Res.* **1991**, *24*, 9–15. [CrossRef]

photonics

MDPI

Article

Angle-Tunable Method for Optimizing Rear Reflectance in Fabry–Perot Interferometers and Its Application in Fiber-Optic Ultrasound Sensing

Yufei Chu [1,2,*], Mohammed Alshammari [1], Xiaoli Wang [1] and Ming Han [1]

[1] Department of Electrical and Computer Engineering, Michigan State University, East Lansing, MI 48825, USA; alsham19@msu.edu (M.A.); wangx250@msu.edu (X.W.); mhan@egr.msu.edu (M.H.)

[2] School of Marine and Atmospheric Sciences, Stony Brook University, Stony Brook, NY 11790, USA

* Correspondence: yufei.chu@stonybrook.edu

Abstract: With the introduction of advanced Fiber Bragg Grating (FBG) technology, Fabry–Pérot (FP) interferometers have become widely used in fiber-optic ultrasound detection. In these applications, the slope of the reflectance is a critical factor influencing detection results. Due to the intensity limitations of the laser source in fiber-optic ultrasound detection, the reflectance of the FBG is generally increased to enhance the signal-to-noise ratio (SNR). However, increasing reflectance can cause the reflectance curve to deviate from a sinusoidal shape, which in turn affects the slope of the reflectance and introduces greater errors. This paper first investigates the relationship between the transmission curve of the FP interferometer and reflectance, with a focus on the errors introduced by simplified assumptions. Further research shows that in sensors with asymmetric reflectance slopes, their transmittance curves deviate significantly from sinusoidal signals. This discrepancy highlights the importance of achieving symmetrical slopes to ensure consistent and accurate detection. To address this issue, this paper proposes an innovative method to adjust the rear-end reflectance of the FP interferometer by combining stress modulation, UV adhesive, and a high-reflectivity metal disk. Additionally, by adjusting the rear-end reflectance to ensure that the transmittance curve approximates a sinusoidal signal, the symmetry of the slope is maintained. Finally, through practical ultrasound testing, by adjusting the incident wavelength to the positions of slope extrema (or zero) at equal intervals, the expected ultrasound signals at extrema (or zero) can be detected. This method converts the problem of approximating a sinusoidal signal into a problem of the slope adjustment of the transmittance curve, making it easier and more direct to determine its impact on detection results. The proposed method not only improves the performance of fiber-optic ultrasound sensors but also reduces costs, paving the way for broader applications in medical diagnostics and structural health monitoring.

Keywords: Fabry–Pérot (FP); fiber Bragg grating; (FPI); fiber-optic ultrasound sensors; reflectance matching

1. Introduction

Fabry–Perot interferometers (FPIs) are highly regarded for their exceptional sensitivity and precision, which form the backbone of numerous applications across both scientific and industrial domains. Their versatility is particularly evident in telecommunications, where FPIs play a crucial role in enhancing signal processing and data transmission over vast networks. By employing tunable optical filters, these devices facilitate dense channel analysis and effectively suppress optical noise, contributing significantly to the efficiency and reliability of modern communication systems [1,2]. This capability has become indispensable in the deployment of large-scale communication infrastructures, driving advancements in the telecommunications industry. In the realm of environmental monitoring, the acute sensitivity of FPIs to optical path variations allows for the precise detection of trace gasses and pollutants, which is critical for environmental protection and pollution control. When integrated with Doppler technology, FPIs can provide wind field data over distances exceeding 100 km, which is vital for climate research and pollution management [3–6]. This application not only enhances our understanding of atmospheric dynamics but also supports sustainable environmental practices.

Moreover, the integration of FPIs with fiber Bragg grating (FBG) technology has led to significant advancements in fiber-optic sensing. By leveraging the stable optical properties within fiber-optics, FPIs enable highly accurate ultrasonic wave detection, thus expanding possibilities in both biomedical and environmental sensing applications [7,8]. These developments have not only improved sensor performance but have also opened up new avenues for high-precision measurements in various fields. Recent innovations include the development of high-sensitivity temperature sensors within FP cavities filled with ultraviolet glue, designed specifically for applications such as real-time ocean temperature monitoring [9–11]. Through continuous efforts, researchers have significantly improved the adaptability and precision of FPIs to maintain high accuracy and reliability in complex and demanding environments. For example, Wang et al. extended fiber-optic FPI technology to Raman lidar for atmospheric water vapor measurements [12,13]. Additionally, Karim et al. expanded the application of modified Phase-Generated Carrier (PGC) demodulation in the field of ultrasound detection [14]. Significant advancements have also been made in fiber-optic FPI temperature measurement techniques [15–17].

FBGs are typically produced through an ultraviolet (UV) light etching process, which causes periodic changes in the fiber's refractive index—essential for the reflective properties required by FP interferometers [15–17]. Increasing the reflectance causes the transmittance curve to gradually deviate from a sinusoidal shape. Although previous studies have shown that when the reflectance is within 26%, this deviation does not significantly impact the results [18], such findings are limited to specific applications. If the reflectivities of the two surfaces of the FPI are not properly selected, it may distort the slope of the reflectance spectrum. Such distortion can result in significant errors in the interferometer's response to ultrasound signals, thereby reducing the effectiveness of the detection system [16,18]. Therefore, a key challenge in optimizing FP interferometers for ultrasonic detection is ensuring that the reflectivities of the FBGs are precisely matched. This requires not only advanced manufacturing techniques but also rigorous testing and calibration. This paper takes a more general approach to FP fiber-optic systems, using the symmetry of the slope of the reflection spectrum as a metric to study the response to ultrasound signals.

This paper explores the fundamental principles of reflectivity in FP interferometers, focusing on how the reflectivity at both the front and back ends impacts the transmission rate and slope. We propose a novel method to adjust the rear reflectivity, which involves embedding a gold disk with UV adhesive at the fiber's rear. By manipulating the stress applied, the angle between the fiber rear face and the gold disk is adjusted, altering the reflectivity accordingly. Experimental results validate the feasibility and effectiveness of this approach. The implications of this research extend the functionality of fiber-optic ultrasonic sensors and, due to its cost-efficiency, broaden their potential applications in fields such as medical diagnostics, structural health monitoring, and non-destructive testing. This paper is structured as follows: Section 2 introduces the reflectance curves and slopes of Fabry–Pérot (FP) interferometers. Section 3 details the method of adjusting the reflectivity at the rear end to develop the FP sensor. Section 4 validates the results of ultrasound detection using the fabricated fiber-optic FP sensor. Finally, Section 5 provides a summary and discussion.

2. Reflectance Curves and Slopes of Fabry–Perot Interferometers

The transmission curve and its slope in FP interferometers are crucial for their performance in various detection applications and analytical tasks. At the core of an FP interferometer are two parallel, highly reflective surfaces. When light enters the interferometer, it undergoes multiple reflections between these surfaces. These repeated reflections cause the light beams to overlap, leading to constructive or destructive interference based on several factors: the wavelength of the light, the angle at which the light enters the interferometer, and the separation between the mirrors, commonly referred to as the cavity length. To provide a clear and intuitive understanding of the transmission characteristics, we can examine a typical FP etalon used in Doppler lidar systems. An FP etalon generally consists of two parallel plates, which are usually made of materials such as glass or quartz. This configuration can also be adapted to feature two mirrors that serve as the reflective surfaces within an FP interferometer. Specifically, in the context of an etalon, the two parallel reflective surfaces are crucial components.

When a plane wave, represented as U_0, strikes the etalon, it undergoes a series of reflections and transmissions between these reflective surfaces. As depicted in Figure 1, the angle of incidence of U_0 is denoted by δ. The amplitude of the transmitted beam after these interactions can then be calculated, providing insights into the behavior of the system under specific conditions [19–21]. Assume the reflectivity of the front surface is R_1, the reflectivity of the rear surface is R_2, the refractive index between the two interfaces is n, and the thickness is h, the angle of incidence is denoted as θ. To visually represent the transmission curve and emphasize the importance of the reflective surfaces within the FP interferometer, this paper starts with the incident light and derives the formula for transmittance as follows [19]:

$$U_0' = U_0(1 - R_1)(1 - R_2) \tag{1}$$

$$U_1' = U_0(1 - R_1)(1 - R_2)R_1R_2e^{i\delta} \tag{2}$$

$$U_2' = U_0(1 - R_1)(1 - R_2)R_1^2R_2^2\,e^{2i\delta} \tag{3}$$

$$U_3' = U_0(1 - R_1)(1 - R_2)R_1^3R_2^3\,e^{3i\delta} \tag{4}$$

where $\delta = \frac{4\pi nhcos\theta}{\lambda}$, and the total amplitude of the transmitted beam is

$$\begin{aligned} T &= \frac{U'U'^*}{U_0U_0^*} \\ &= \frac{(1-R_1)^2(1-R_2)^2}{(1-R_1R_2)^2+4R_1R_2sin^2\left(\frac{\delta}{2}\right)} \end{aligned} \tag{5}$$

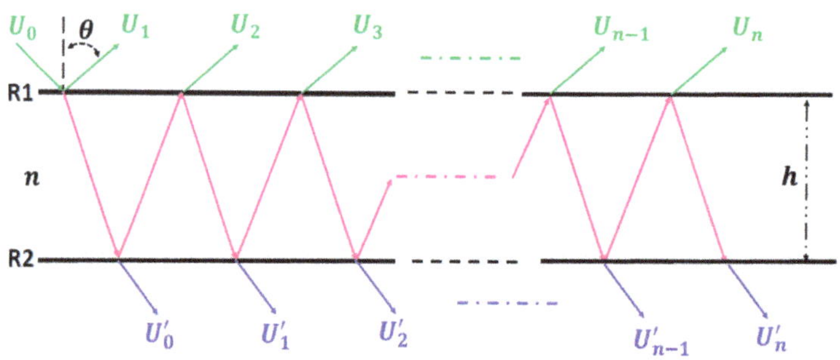

Figure 1. The schematic diagram of the FP transmittance curve.

Assuming that the reflectivity is low, and the amplitude reflectance of the two faces of the etalon are equal, that is, $R1 = R2$, the intensity reflectance of each face is as follows: $R = Ri^2$. In most current FBG applications, for the sake of computational simplicity, we typically assume that the reflectivity is very low and that the reflectivity of the two surfaces is identical. In this case, $(1 - Ri)^2$ is simplified to $(1 - Ri^2)$. In most applications, this simplification is reasonable and does not significantly affect the measurement results. However, in certain precise measurements, such as ultrasonic detection, this simplification can introduce errors. In practical applications, to enhance the detection signal without increasing the demand for higher incident laser power, people can increase the reflectivity of the sensor. However, while this increases the signal strength, it can also affect the quality of the signal. Thus, it is crucial to understand how the relative error changes with reflectivity, as illustrated in Figure 2, which depicts the variation of errors in $(1 - Ri)^2$ and $(1 - Ri^2)$ with respect to reflectivity.

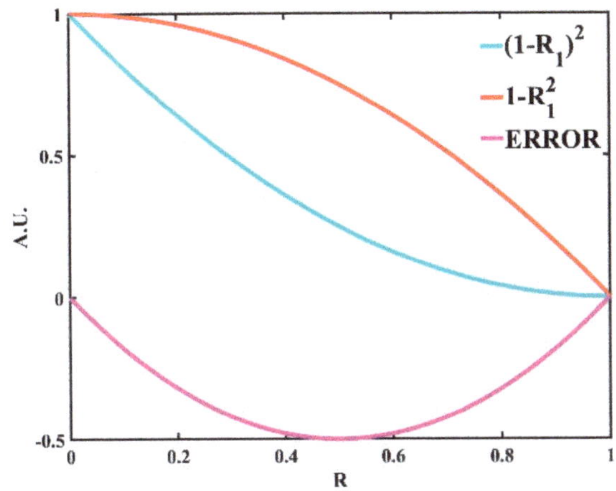

Figure 2. Variation of error due to assumptions in FP reflectivity changes.

As shown in Figure 2, the difference between the expected reflectivity and the measured reflectivity increases as reflectivity rises, reaching a peak at 50%, and then decreases as reflectivity continues to increase. This observed error trend must be considered in practical applications. Fortunately, in most applications, where reflectivity is less than 5%, the relative error, given by $1 - 0.05^2 - (1 - 0.05)^2 = 0.095$, is negligible and can be ignored. In practical applications, understanding the transmission curve of the FP interferometer allows us to determine whether the errors are within an acceptable range for detection when the FP is used as a filter, frequency discriminator, etc. [20].

In the field of fiber-optic ultrasonic detection, improving sensitivity is a key objective. To this end, the Han group pioneered a coiled fiber configuration that significantly enhances the signal-to-noise ratio (SNR) [14,22]. This innovative setup requires the use of longer FP cavity lengths, which reduce the full width at half maximum (FWHM) and necessitate highly stable laser wavelengths for optimal performance. The most common approach involves laser frequency stabilization; however, this method is typically expensive, limiting its application range. To address this, Han and colleagues developed additional strategies to mitigate the impact of laser wavelength drift, ensuring that ultrasonic detection remains stable and reliable even as laser wavelengths fluctuate with environmental changes [14,22]. These strategies are based on the assumption that $(1 - Ri)^2 = (1 - Ri^2)$. However, as shown in Figure 2, these assumptions may introduce a certain degree of error. The response in ultrasound detection is not only related to Figure 2 but also relies on a fundamental assumption that the transmittance curve is sinusoidal [14,18]. While it is generally assumed that lower reflectance is better, there is no specific metric to evaluate the extent of error caused by deviations in the reflectance curve. Therefore, this paper uses the slope of the reflectance to visually represent the error and further investigates how the slope affects the detection results.

To minimize the impact of this error, this paper conducts an in-depth study of the principles behind FBG ultrasound detection. When ultrasound waves pass through the FBG sensor, they cause slight changes in the refractive index, resulting in a phase shift (or a change in the relative position of the laser wavelength to the transmission curve). The detection sensitivity is directly related to the slope; the steeper the slope, the higher the sensitivity. The actual detected signal is the product of the reflectance curve and the reflected light intensity. For simplification, this discussion initially considers only the intensity changes caused by slope variations.

However, with changes in reflectance, the assumption of a sinusoidal shape for the transmittance curve may no longer hold. As shown in Figure 3, variations in reflectance can lead to significant deviations of the transmittance curve from a sinusoidal shape, thereby affecting the accuracy and reliability of the detection process. In Figure 3a, when $R1 = R2 = 1\%$, the reflectance curve is nearly sinusoidal, and its slope is almost symmetrical. However, when the reflectance increases to 30%, it is evident that the reflectance curve is no longer sinusoidal, and its slope is no longer symmetrical. As the reflectance continues to rise, the asymmetry of the slope and the deviation of the transmittance curve from the sinusoidal shape become more pronounced. It is important to note that in Figure 3, for ease of comparison of slopes at different reflectance levels, each transmittance curve and its corresponding slope have been normalized by dividing by their respective maximum values. This normalization allows for a clearer comparison of how the shape and steepness of the curves change with varying levels of reflectance. Since the slope of the reflectance more directly impacts the detection signal, we will subsequently focus on the slope of the transmittance curve.

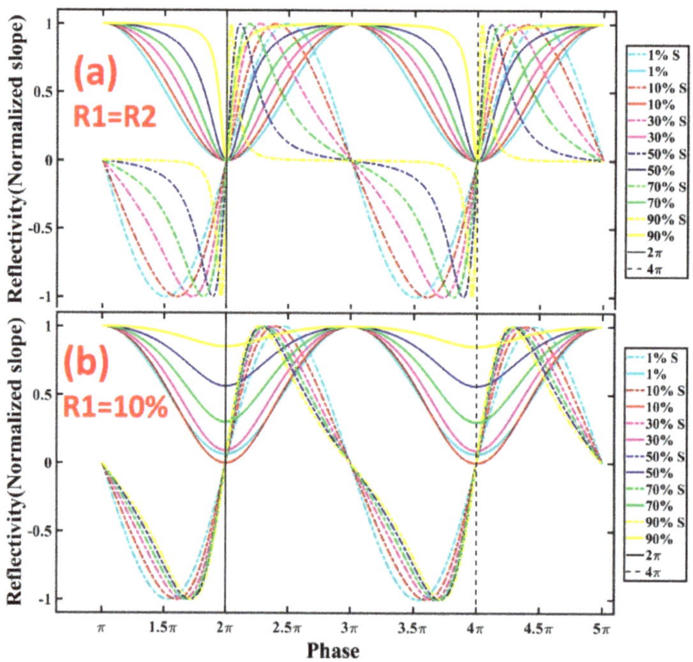

Figure 3. The variation in the slope due to the FP reflectivity changes. (**a**) The reflectivity of the front and rear ends is equal; (**b**) the reflectivity of the front end is fixed at 10%, and only the reflectivity of the rear end is changed; the solid line is the reflectivity; the dotted line is the slope of the normalized reflectivity. In the legend, "S" represents the slope.

When $R1$ and $R2$ are not equal, the slope of the reflectivity curve increases with the difference between the reflectivities. As shown in Figure 3b, we fixed $R1 = 10\%$ while varying $R2$. When $R1$ has a reflectivity of 10%, the optimal reflectivity of $R2$ is determined by the application scenario. In general applications, choosing R2 with a reflectivity of 10% is preferable. However, when considering the introduction of an additional light path or phase-modulated incident light to mitigate wavelength drift, as in the phase-generated carrier demodulation scheme in [18], it is necessary to assume that the reflectivity curve is sinusoidal, with a symmetric slope. If $R2$ is set to 10%, the reflectivity curve may deviate from the sinusoidal shape, and the asymmetry of the slope can clearly indicate the extent of deviation. By comparing the slope curves in Figure 3b, we can see that choosing R2 at 1% is more suitable in this context. The figure demonstrates that when $R2 = 1\%$, the reflectivity slope is most symmetric. However, as $R2$ increases, the slope becomes increasingly asymmetric. It is important to note that in Figure 3a, where $R1 = R2$, all the reflectivity curves reach the minimum of zero. However, as illustrated in Figure 3b, the reflectivity curve only reaches zero when both $R1$ and $R2$ are set to 10%; in all other cases, it does not. This characteristic is particularly advantageous for certain practical applications, such as when the backend detection component includes an envelope detector. Since it requires a specific operating threshold, it is preferable that the reflectivity does not drop to zero.

Based on the above analysis, it is clear that in ultrasound detection applications using FP sensors, such as those employing enhanced phase-generated carrier demodulation techniques [14], it is essential to assume that the reflectance curve is sinusoidal, with the transmittance slope being approximately linear and symmetrical. Satisfying this assumption is crucial to ensuring that one laser wavelength is positioned at the midpoint of the slope while another wavelength is at the peak or trough, regardless of frequency (phase) shifts. Symmetrical slopes are the foundation for maintaining the accuracy of this setup. However, as shown in Figure 3, deviations from these ideal conditions often occur. Variations in reflectance between the front and rear surfaces of the sensor can lead to changes in the transmittance curve and its slope. Specifically, as illustrated in Figure 3a, lower reflectance tends to result in more symmetrical slopes. However, achieving lower reflectance requires the use of higher laser energy. Figure 3b further demonstrates that the mismatch in reflectivity between the front and rear reflective surfaces of the FP (due to different reflectivities) results in distinct overall reflection curves and slopes.

3. Angle Adjustment of End-Face Reflectivity Based on UV Glue and Gold Disk

The results discussed in the previous section require the transmittance curve to have symmetrical slopes. According to the findings in Figure 3, this necessitates precise matching of the reflectance between the front and rear surfaces of the FP sensor. If only R1 uses the FPI, modifying $R2$ from an FPI to a gold disk can be used to adjust the reflectivity of $R2$. This is achieved by altering the angle of $R2$ (gold disk) to control the proportion of light reflected into the fiber, thereby adjusting $R2$'s reflectivity. The reflectivity curves at different $R2$ settings are illustrated by the solid line in Figure 3b. Note that the final angle should be determined based on real-time monitoring of the reflectivity curve. Adjusting the reflectance of the FPI typically involves techniques such as XCL UV etching. However, as the demand for higher precision increases, so does the cost [20]. To accurately control reflectance while reducing manufacturing costs, we introduce an innovative method. Initially, the front FPI is created using XCL UV etching to establish $R1$. Then, depending on specific detection requirements, the reflectance $R2$ is adjusted at a specific length behind R1 to meet these requirements. This is achieved by fixing a gold disk (or other high-reflectivity material) at the fiber end with UV adhesive and adjusting the angle between the gold disk and the fiber end before UV curing. The detailed manufacturing process is as follows:

(1) Initial FPI creation: Using XCL UV etching, a front FPI is fabricated with the desired reflectivity R1. The fiber is positioned within the XCL optical path, and by carefully controlling the power and exposure time of the XCL laser, the reflectivity R1 can be precisely adjusted. It is important to note that the outcomes of this process can vary depending on the specific XCL optical setup and the material properties of the fiber. Therefore, after laser etching, it is essential to accurately measure and record the reflectivity of the FPI. This ensures that the fabricated FPI meets the required specifications and allows for any necessary adjustments in subsequent manufacturing steps [17].

(2) The preparation of the end side of fiber: Based on the specific requirements of the sensor, the fiber end face is carefully polished at a precise location behind R1. This ensures a smooth and flat surface that is essential for achieving the desired reflectivity. The fiber is then securely positioned on a three-dimensional translation stage using a fiber holder. This setup allows for precise alignment and positioning of the fiber end face, which is critical for the subsequent attachment of the reflective material. The accurate positioning ensures that the final reflectivity meets the specific sensor requirements, thereby optimizing the sensor's overall performance.
(3) Alignment: Place the gold disk on a three-dimensional moving platform and align the fiber using electron microscopy to ensure precise and gentle contact with the gold disk. The gold disk (reflector) needs to be larger than the fiber diameter but not excessively so to prevent detachment. As a compromise, we selected a reflector with a diameter of 600 μm.
(4) Reflectivity adjustment: Apply 0.5 mL of UV glue to the center of the gold disk using a syringe or micropump, then adjust the stress to modify the fiber's angle relative to the disk. Continuously monitor the transmission rate curve and fine-tune the stress (angle) to achieve the desired specifications. The adjustment range of the gold disk (reflector) is theoretically from 0 to the reflectivity of the gold disk (~90%), which fully meets our required range of adjustment.
(5) UV curing: solidify the configuration using a UV lamp to permanently set the angle between the fiber and the gold disk [16].

The results demonstrate that by continuously adjusting the angle, the sensor's slope can be effectively fine-tuned. When the slope becomes symmetrical, the reflectance curve closely approximates a sinusoidal signal, making it suitable for subsequent ultrasound detection. Figure 4a,b present the experimentally measured reflectance values and their corresponding slopes. In Figure 4a (highlighted in orange), the slope exhibits asymmetry, indicating that the reflectance curve can no longer be approximated as sinusoidal. In contrast, the slope (orange) in Figure 4b shows a high degree of symmetry, and the reflectance curve can be considered approximately sinusoidal. At this point, the FP sensor with symmetrical slopes can be used for ultrasound detection, which is applicable in phase-generated carrier (PGC) demodulation systems [14] or other FP-based detection applications that require a transmittance curve approximating a sinusoidal shape [23].

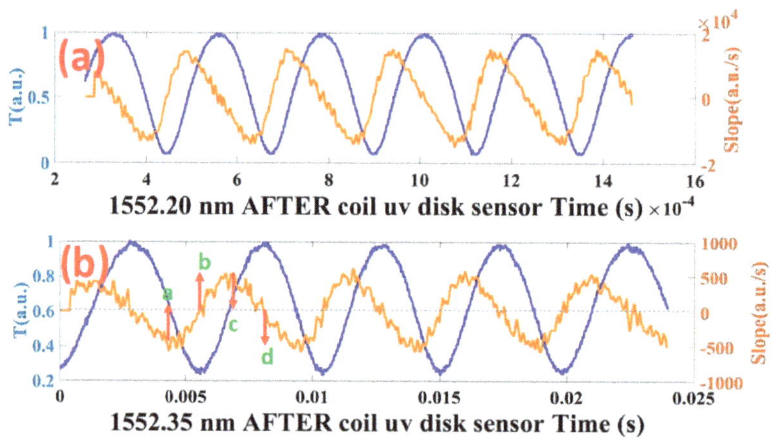

Figure 4. The angle adjustment of end–face reflectivity based on UV Glue. (**a**) The slope of the reflectivity is asymmetric; (**b**) the slope of the reflectivity is symmetric (the green points labeled a, b, c, and d correspond to the positions where the slope is at a minimum, zero, maximum, and zero, respectively).

Using a high-reflectivity metal disk at the back end can significantly expand the usable wavelength range of the fiber-optic FP sensor. When the transmittance of both the front FPI and the back-end FPI varies with wavelength, the range of incident laser wavelengths that meet detection requirements is greatly reduced. However, in the method proposed in this paper, the back-end reflectance can be approximately considered independent of wavelength, meaning only the front-end FPI varies with wavelength. This approach can significantly expand the working range of laser wavelengths. In the experiment, we used a piezoelectric transducer to precisely adjust the incident laser wavelength. We observed that since the reflectance of the front FPI varies with wavelength, this corresponds to R2 being fixed while R1 varies with wavelength, as shown in Figure 4. Figure 4a,b illustrate the changes in reflectance and its slope with varying laser wavelengths.

4. Results

To evaluate the actual performance of the FP ultrasound sensor, an FP ultrasound sensing system was set up as shown in Figure 5a. Both the fiber-optic FP sensor and the piezoelectric transducer were mounted on an aluminum plate, 25 cm apart. We utilized a Chirped Fiber Bragg Grating (CFBG) [17] to extend the operating wavelength of the first reflective surface to 1550 $\pm$ 5 nm. This system is designed to detect ultrasound waves generated by a piezoelectric transducer driven by a waveform generator, using the aforementioned fiber-optic FP sensor. The laser beam from the laser source is connected to port 1 of the circulator, with port 2 connected to the FP ultrasound sensor. The collected signal is then amplified by 40 dB using a photodetector (NEWPORT 1811-FC) connected to port 3 of the circulator, followed by a low-pass filter (500 kHz), and then fed into an oscilloscope (OSC, RIGOL DS1104B) or data acquisition card (DAQ). The fiber-optic FP sensor has a length of 38.97 cm and is coiled between R1 and R2 to form a disk with an inner diameter of approximately 6 mm and an outer diameter of approximately 8 mm [18]. The FP sensor is then fixed onto the aluminum plate using glue. The waveform generator produces a 150 kHz, 5 V peak-to-peak, five-cycle burst sinusoidal signal with a repetition rate of 40 Hz to drive the piezoelectric transducer and generate ultrasound waves.

The results demonstrate that sensors with good slope symmetry achieve the expected experimental outcomes. Figure 4a,b show the actual measured reflectance values and their corresponding slopes. It is evident that the slope in Figure 4a (indicated in orange) exhibits asymmetry, while the slope in Figure 4b shows a high degree of symmetry. For wphen the incident wavelength is adjusted to positions a, b, c, and d in Figure 4b, the corresponding results are shown in Figure 5b–e. When the laser wavelength is at the minimum value (position a), the detected signal strength is at a maximum. Similarly, when the laser wavelength is at the maximum value (position c), the detected signal strength is also at a maximum. When the laser wavelength is at points where the slope is zero (positions b and d), the ultrasound signal is zero. These experimental results exhibit strong symmetry. The four laser wavelengths in Figure 4b are equally spaced, with the slope exhibiting odd symmetry at points b and d, and even symmetry at points a and c. However, when using the sensor shown in Figure 4a, the intervals between the maxima and minima lack any symmetry.

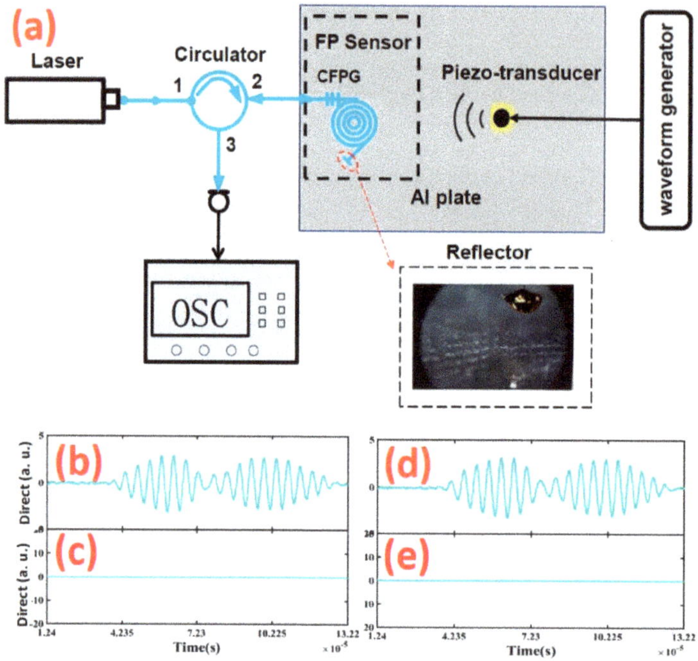

Figure 5. Angle symmetry testing setup and results for the ultrasound sensor: (**a**) ultrasound detection setup (the structure diagram of the FP sensor is shown within the black dashed line), (**b–e**) show the ultrasound detection results when the laser wavelength is adjusted to operating points a, b, c, and d in Figure 4b, respectively.

5. Conclusions and Discussion

This paper investigates the relationship between the transmission curve of the FP interferometer and reflectance, identifying potential errors introduced by simplified assumptions. It further examines the effects of asymmetrical slopes when the assumption of a sinusoidal transmittance curve is not met, which can significantly influence detection accuracy. To address this issue and achieve symmetrical slopes in the transmittance curve, this study proposes an innovative method for adjusting the back-end reflectance of the FP interferometer. This is accomplished through a combination of stress modulation, UV adhesive, and a high-reflectivity metal disk (gold disk). The effectiveness of the proposed method was validated through practical ultrasound detection experiments, which demonstrated improved symmetry in the detection results. This research offers a more cost-effective detection approach, potentially expanding the applicability of this technology. The results indicate that the detected signal strength reaches its maximum when the laser wavelength is at the slope's minimum (position a) and maximum (position c), while the ultrasound signal is zero at points where the slope of the transmittance curve is zero (positions b and d).

This paper comprehensively investigates the relationship between reflectance, transmittance, and the slope of the FP transmission curve, and introduces a cost-effective method for adjusting the end-face reflectance by modifying the angle between the fiber end face and a metal disk. Although a metal (gold) disk was used in this study, other similar reflective materials could potentially yield comparable results. The back-end reflectance is not wavelength-selective, which significantly broadens the usable wavelength range of the laser. The next step is to apply this FP sensor with the symmetric slope to practical scenarios where environmental interference can be mitigated (e.g., using a PGC scheme with an electro-optic modulator (EOM) or a dual-path approach with an acousto-optic modulator (AOM)) and to further compare its SNR performance. Future work could explore fixing the angle between the metal disk and the fiber end face, and utilizing a CO_2 laser or femtosecond laser to etch the fiber to achieve the desired reflectance. This approach offers an additional means for fine-tuning reflectance, further enhancing the flexibility and applicability of sensor technology. The degree to which the reflectivity–transmittance curve deviates from a sinusoidal shape is often difficult to assess intuitively. However, in this paper, we propose using the slope of the reflectivity as a more intuitive metric to evaluate the extent of this deviation. This method can also serve as an indicator for evaluating how much the FP transmittance curve deviates from a sinusoidal shape in subsequent analyses. While this method was initially developed to adjust reflectance based on slope symmetry, it is equally applicable to general fiber-optic FP sensors without symmetry requirements [23], offering the potential to substantially reduce manufacturing costs while expanding the operating wavelength range.

Author Contributions: Conceptualization, Y.C.; methodology, Y.C.; software, Y.C.; data curation, Y.C. and X.W.; writing—original draft preparation, Y.C. and M.A.; writing—review and editing, M.H.; visualization, Y.C.; supervision, M.H.; project administration, M.H.; funding acquisition, M.H. All authors have read and agreed to the published version of the manuscript.

Funding: This research was funded by the Office of Naval Research (N000142112273, N000142212321).

Institutional Review Board Statement: Not applicable.

Informed Consent Statement: Not applicable.

Data Availability Statement: The data underlying the results presented in this paper are not publicly available at this time but may be obtained from the authors upon reasonable request.

Acknowledgments: Y. Chu would like to thank Hasanur Rahman Chowdhury for his help with the sensor fabrication and Qiwen Sheng for the technical discussion on FP modeling.

Conflicts of Interest: The authors declare no conflicts of interest.

References

1. Javernik, A.; Donlagic, D. High-speed interrogation of low-finesse Fabry–Perot sensors using a telecom DFB laser diode. *J. Light. Technol.* **2017**, *35*, 2161–2171. [CrossRef]
2. Moradi, H.; Zhoulideh, M.; Ghafariasl, M. Tunable and ultrasensitive sensor covering terahertz to telecommunication range based on a Fabry–Perot interference of graphene plasmonic waves. *Opt. Commun.* **2023**, *542*, 129592. [CrossRef]
3. Tan, L.; Hua, D.; Wang, L.; He, T.; Wang, Y.; Xing, M. Performance comparison of Fabry–Perot and Mach–Zehnder interferometers for molecular Doppler lidar based on fringe-imaging technique. *Optik* **2014**, *125*, 6291–6295. [CrossRef]
4. Chu, Y.; Liu, D.; Wang, Z. Basic principle and technical progress of Doppler wind lidar. *Chin. J. Quantum Electron.* **2020**, *37*, 580.
5. Chu, X.; Chen, Y.; Cullens, C.Y.; Yu, Z.; Xu, Z.; Zhang, S.R.; Huang, W.; Jandreau, J.; Immel, T.J.; Richmond, A.D. Mid-latitude thermosphere-ionosphere Na (TINA) layers observed with high-sensitivity Na Doppler lidar over Boulder (40.13° N, 105.24° W). *Geophys. Res. Lett.* **2021**, *48*, 2021GL093729. [CrossRef]
6. Zhao, M.; Chen, J.; Xie, C.; Li, L. The impacts of deformed Fabry–Perot interferometer transmission spectrum on wind lidar measurements. *Remote Sens.* **2024**, *16*, 1076. [CrossRef]
7. Domingues, M.F.; Tavares, C.; Alberto, N.; Radwan, A.; André, P.; Antunes, P. High rate dynamic monitoring with Fabry–Perot interferometric sensors: An alternative interrogation technique targeting biomedical applications. *Sensors* **2019**, *19*, 4744. [CrossRef] [PubMed]
8. Liu, B.; Li, Y.; Wang, R.; Chen, X.; Li, J.; Chen, H.; Jiang, M. Label-free and selective heparin detection by surface functionalized fiber Fabry-Perot interferometer biosensor. *Opt. Fiber Technol.* **2024**, *84*, 103770. [CrossRef]
9. Zhang, S.; Mei, Y.; Xia, T.; Cao, Z.; Liu, Z.; Li, Z. Simultaneous measurement of temperature and pressure based on Fabry-Perot interferometry for marine monitoring. *Sensors* **2022**, *22*, 4979. [CrossRef] [PubMed]
10. Han, G.L.M.; Hou, W. Fiber-Optic Temperature and Flow Sensor System and Methods. Patent US9995628B1, 12 June 2018. Available online: https://patents.google.com/patent/US9995628B1/en (accessed on 27 October 2024).
11. Liu, G.; Han, M.; Hou, W.; Matt, S.; Goode, W. A miniature fiber-optic sensor for high-resolution and high-speed temperature sensing in ocean environment. *Proc. SPIE* **2016**, *9459*, 80–85.
12. Wang, Y.; Jia, L.; Li, X.; Fan, F.; Di, H.; Song, Y.; Hua, D. Compact fiber-optic spectroscopic design and its validation in atmospheric water vapor Raman lidar. *J. Opt. Soc. Am. B* **2020**, *37*, 384944. [CrossRef]
13. Wang, Y.; Wang, Q.; Hua, H.D. Preliminary exploration of atmospheric water vapor, liquid water, and ice water by ultraviolet Raman lidar. *Opt. Express* **2019**, *27*, 36311–36321.

14. Karim, F.; Zhu, Y.; Han, M. Modified phase-generated carrier demodulation of fiber-optic interferometric ultrasound sensors. *Opt. Express* **2021**, *29*, 432237. [CrossRef]
15. Uddin, N.; Sheng, Q.; Lee, S.; Reinke, M.L.; Donovan, D.; Shafer, M.; Han, M. Multichannel fiber-optic silicon Fabry–Pérot interferometric bolometer system for plasma radiation measurements. *Photonics* **2021**, *8*, 344. [CrossRef]
16. Jiang, X.; Duan, Z.; Xu, T.; Mupfukirei, L.R.; Yu, Z.; Pang, B.; Wang, Y.; Ling, Q.; Chen, H.; Gu, Z.; et al. Review of fabrication and packaging of UV-induced FBGs for high temperature sensing. *Opt. Fiber Technol.* **2024**, *86*, 103855. [CrossRef]
17. Chowdhury, H.R. Fiber Optic Fabry-Pérot Interferometric Sensor for Temperature and Strain Measurement. Ph.D. Thesis, Michigan State University, East Lansing, MI, USA, 2023.
18. Karim, F.; Han, M. Effect of three-beam interference on phase-generated carrier demodulation of a fiber-optic interferometric sensor. *Opt. Express* **2023**, *31*, 9769–9778. [CrossRef] [PubMed]
19. Wang, Y.; Chu, Y.; Liu, D.; Wu, D.; Zhenzhu, W.A.; Kunming, X.I.; Kuang, Z.; Bangxin, W.A.; Zhong, Z.; Aiyuan, F.A.; et al. Device for Measuring Transmittance Curve of Fabry-Perot Using Frequency Comb Light Source and Method Using the Same. Patent US11874169B2, 16 January 2024. Available online: https://patents.google.com/patent/US11874169B2/en (accessed on 27 October 2024).
20. Sengupta, D. *Optical Fiber Sensors*; CRC Press: Boca Raton, FL, USA, 2015.
21. Chu, Y. Research on Technology and Method of Detecting Boundary Layer Parameters Based on Various Lidars. Ph.D. Thesis, University of Science and Technology of China, Hefei, China, 2020.
22. Liu, G.; Zhu, Y.; Sheng, Q.; Han, M. Polarization-insensitive, omnidirectional fiber-optic ultrasonic sensor with quadrature demodulation. *Opt. Lett.* **2020**, *45*, 397955. [CrossRef] [PubMed]
23. Alshammari, M.; Chu, Y.; Han, M. Real-time in-situ phase sensitivity calibration of interferometric fiber-optic ultrasonic sensors. *Opt. Lett.* **2024**, *49*, 5336–5339. [CrossRef] [PubMed]

photonics

MDPI

Article

Research and Application of Interferogram Acquisition Method for Ground-Based Fourier-Transform Infrared Greenhouse Gas Spectrometer

Yasong Deng [1,2], Liang Xu [1,*], Ling Jin [1], Yongfeng Sun [1], Lei Zhang [1,2], Jianguo Liu [1] and Wenqing Liu [1]

1 Key Laboratory of Environmental Optics and Technology, Anhui Institute of Optics and Fine Mechanics, Hefei Institutes of Physical Science, Chinese Academy of Sciences, Hefei 230031, China; ysdeng@aiofm.ac.cn (Y.D.)

2 School of Environmental Science and Optoelectronic Technology, University of Science and Technology of China, Hefei 230026, China

* Correspondence: xuliang@aiofm.ac.cn

Abstract: A high-performance data acquisition and processing system within a spectrometer provides a powerful guarantee for obtaining high-precision data in ground-based Fourier-transform infrared greenhouse-gas spectroscopy. Addressing the challenge of accurate interferogram sampling in Fourier-transform spectroscopy, a dual-channel interferogram acquisition method was designed. Dual-channel analog-to-digital converters, acquiring interferograms at different gains, enable high dynamic range and high-resolution acquisition of infrared interferometric signals; the analog-to-digital converter channel of low-gain interferograms mainly captures data near the zero-optical-range-difference spike, and the analog-to-digital converter channel of high-gain interferograms mainly acquires the weak signals from the two flanks. The simulation results, circuit design, and correction method between the two channels of the method are given. Finally, in the ground-based Fourier-transform infrared greenhouse-gas spectrometer for experimental applications, the experiment shows that under the same measurement conditions, the carbon dioxide column concentration-measurement accuracy is improved by 2.096 times, and the dual-channel interferometric data acquisition method can significantly enhance the data retrieval accuracy.

Keywords: Fourier-transform infrared spectroscopy; interferogram acquisition; greenhouse gases; carbon dioxide

1. Introduction

Ground-based Fourier-transform infrared (FTIR) measurements are available at various locations around the world for determining atmospheric trace gas concentrations and investigating climate change [1–3]. Fourier-transform spectrometers have become the tool of choice in the field of high-performance spectral analysis because of their excellent performance in simultaneous signal processing, high-throughput, high resolution, and high wavelength accuracy [4,5]. As the core of FTIR spectroscopy system, the Michelson interferometer is used to obtain spectra by measuring interferograms and performing Fourier transforms to characterize the morphology and quantify the substances [6]. However, sampling errors in interferograms are an important source of spectral noise in conventional sampling methods [7]. Any error in the sampling process will have a significant impact on the spectral performance, which is manifested in the appearance of spectral artifacts, shift in the spectral position, and increase in the spectral noise [8]. These effects not only reduce the clarity and resolution of the spectrum, but also seriously interfere with the accuracy and reliability of spectral analysis [9]. Therefore, in the application of the FTIR spectrometer, how to accurately control the sampling process of the interferogram and reduce the sampling error is key to improving the spectral performance and ensuring the accuracy of the spectral analysis. Furthermore, the results also highlight new challenges and requirements for the design and optimization of spectroscopic instruments.

Inaccurate sampling of the interferogram may result in the generation of another type of noise in the spectrum, known as quantization or digitization noise [10]. The source of digitization noise is illustrated in Figure 1, where a discrepancy between the sampled and true values of the waveform is evident. This discrepancy corresponds to noise in the signal and will translate into spectral noise. In Fourier-transform spectroscopy, an analog time-domain signal is sampled at equal intervals and then subjected to a discrete Fourier transform to obtain a discrete frequency-domain spectrum. Even when the time-domain analog signal is noiseless, rounding errors during the sampling process introduce quantization "noise" [11].

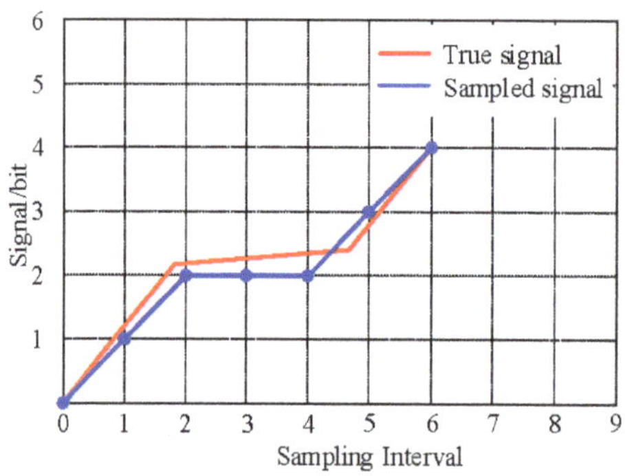

Figure 1. Sources of digitized noise. The difference between the actual signal and the sampled signal at each data point is demonstrated. This error can be approximately considered equivalent to the peak-to-peak detector noise of 1/2 bit.

Traditional interferogram data acquisition methods primarily rely on single-channel analog-to-digital converters (ADCs). Although this method fulfills basic acquisition needs to a certain extent, its inherent limitations, such as sampling errors, quantization noise, and limited dynamic range, severely hinder the further enhancement of spectral analysis performance [12]. These limitations are particularly evident when dealing with spectral analysis of wide spectral ranges or complex substances. To address the issue of interferogram sampling, researchers have proposed various improved methods from different fields. One study [13] established a theoretical model linking the relative error of spectral measurement to the deviation of sampling points for the analysis and correction of spectrometer sampling errors based on Fourier inverse-transform and error-analysis methods. One study [7] introduced analog and digital amplitude and phase components as compensated acquisition systems to mitigate spectral noise caused by sampling errors. However, excessive compensation also increases system uncertainty.

To overcome these challenges, researchers have begun exploring the application of dual-channel ADCs in interferogram data acquisition. By acquiring interferometric signals simultaneously from two or more channels, the dual-channel analog-to-digital converter (ADC) technique not only effectively increases the sampling rate and resolution, but also significantly reduces quantization noise, thereby improving the overall quality of spectral data. Furthermore, the dual-channel acquisition strategy aids in achieving more accurate spectral recovery and reducing the occurrence of spectral position shifts and spectral ghosting, which is crucial for enhancing the accuracy and reliability of spectral analysis. One study [14] designed an acquisition system based on segmented gain control for an FTIR spectrometer, where the infrared interferometric signal peaks at zero optical-path difference (ZPD). The signal attenuation outside the ZPD is similar to that of noise, exhibiting a large overall dynamic range. The acquisition system employs segmented gain control, utilizing a programmed gain amplifier module and an Field Programmable Gate Array (FPGA)-based multiplexed acquisition module to acquire infrared interference signals with a high dynamic range. Simulated interference signal tests demonstrate that this design enhances weak signal acquisition outside the ZPD, improves system dynamic range, and accurately restores the real spectrum.

Regarding the above issue, the interferogram may exhibit an abnormally high signal-to-noise ratio (SNR) due to its inherent properties, particularly at zero delay. As the interferogram region moves away from the ZPD, the phase difference of the waves increases, weakening the amplitude superposition effect and resulting in a gradual decrease in signal strength. Simultaneously, background noise becomes more prominent, causing the SNR of the interferogram region away from the zero-delay point to decrease significantly. In this paper, we propose a dual-channel interferometric data-acquisition method that simultaneously acquires interferometric data near and far from the zero-delay point. This method leverages information from the high-SNR region, while reducing the interference from the low-SNR region on overall measurement results.

In Section 2, we present the theory and specific implementation steps of the dual-channel interferometric data-acquisition method. In Section 3, we conduct a simulation analysis of the acquisition method and present the hardware design of the acquisition circuit based on MATLAB(2016b) software. Additionally, we provide methods for calibrating and correcting errors between the dual channels. In Section 4, we apply the proposed dual-channel interferometric data-acquisition method to a ground-based greenhouse-gas spectrometer to obtain observation results for carbon dioxide (CO_2) column concentration. Finally, some conclusions are given in Section 5.

2. Dual-Channel Interferometric Data-Acquisition Methods

In order to more thoroughly sample the signals on both flanks of the IR interferogram, more and more instruments are using sampling of ADC gain switching to increase the dynamic range of interferogram sampling. As shown in Figure 2, the gain is switched once for $g = 2^N$ (N is an integer) data points on both sides of the ZPD position, and the gain at positions away from the ZPD is usually amplified by more than 8 times that at the ZPD position. Therefore, the principal ZPD position data are amplified to obtain more accurate sampling.

In an FTIR spectrometer, the accuracy of the ADC must exceed the dynamic range in order to discriminate the smallest signal elements in the range. When the digital voltage signal or the ADC's dynamic range is insufficient, certain sharp spectral signals in the interferogram may become blurred, resulting in the useful interferogram signals being obscured by noise. In cases where the acquisition ADC's accuracy is limited, in order to improve the accuracy of the weak signals in the two wings of the interferogram, according to the characteristics of the interferogram, the design employs a segmented variable-gain method: a low gain is used at the zero optical-path difference spike, and, as the signal diminishes, it switches to a high gain, which amplifies the small signals in the two wings, thereby enhancing quantization efficiency during sampling and improving sampling accuracy. When the gain is applied, the signal is quantized and captured by the ADC, resulting in an equivalent increase in resolution by g bits.

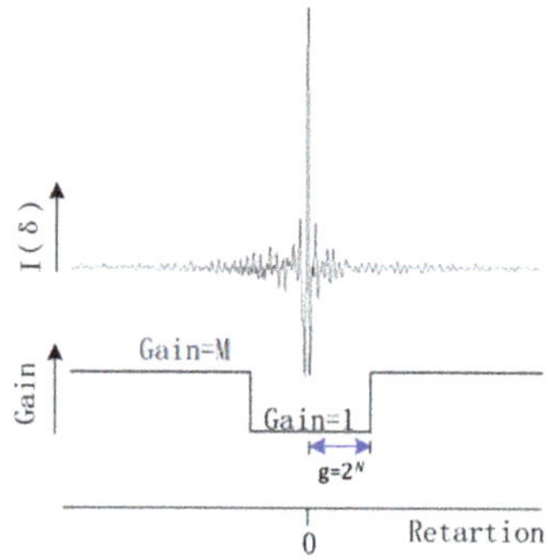

Figure 2. Schematic diagram of amplified gain range.

When the gain is minimized, the ADC's dynamic range must be at least as large as the interferogram's dynamic range. When variable gain amplification is performed on small signals, the amplified signal intensity should not exceed the intensity of the zero optical-range difference spike, lest it exceed the ADC's dynamic range. Maximum gain can be applied to signals whose intensity is close to the RMS value. However, gain switching introduces complexities in precise synchronization control and may cause data fluctuations due to unstable control of the gain switching process. Therefore, a dual-channel interferometric data acquisition method is designed to achieve high-accuracy interferogram sampling with a high dynamic range.

The dual-channel interferometric data acquisition method primarily consists of dual-channel ADC acquisition and low-delay square-wave triggering. Moreover, the system block diagram illustrating this method is shown in Figure 3. Based on the working characteristics of the interferometer, the gain adjustment process involves dividing the detection signal into two channels, firstly, and then the signals of each channel are amplified with different gains, respectively, while the two channels remain parallel and independent of each other.

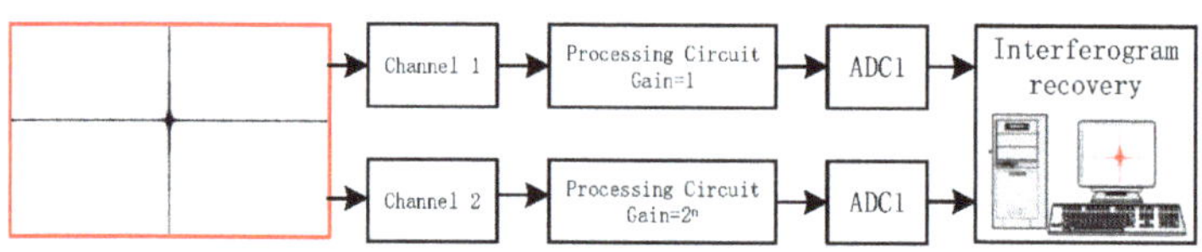

Figure 3. Schematic of data acquisition process.

After obtaining two channels of interferometric data, it is also necessary to carry out the normal calibration procedure on the interferograms. Using open sources or spectrometers for vehicle applications, the interferograms will show jitter, affecting the recovery of the interferograms. Each data point on the calibrated two-channel interferogram needs to be divided by the corresponding gain magnification, so as to convert the entire interferogram to the same effective gain before the Fourier transform. This process of recovering the true interferometric signal is also referred to as data synthesis. The details of the data synthesis process, which combines two channels of data into a single set through splicing, are given in Section 3.1.

3. Simulation Analysis and Circuit Design

3.1. Simulation Experiments and Analysis

A dual-channel 16-bit ADC acquisition card with a dynamic range of −0.1 to 0.1 was simulated using MATLAB software (R2016b). A raw interferogram, used as the data input for ADC channel one, was processed without magnification (Figure 4a). The same interferogram, magnified by a factor of 8, served as the data input for ADC channel two (Figure 4b). Figure 4c,d show the quantization results of the two channels, respectively. Since channel two is primarily designed to capture small signals from the interferogram wings, gain amplification is often applied, but the amplified signal intensity must not exceed the zero optical-range difference spike intensity, otherwise, it will exceed the ADC's dynamic range. The center region of the ZPD in channel two's interferogram approaches the upper limit of this dynamic range.

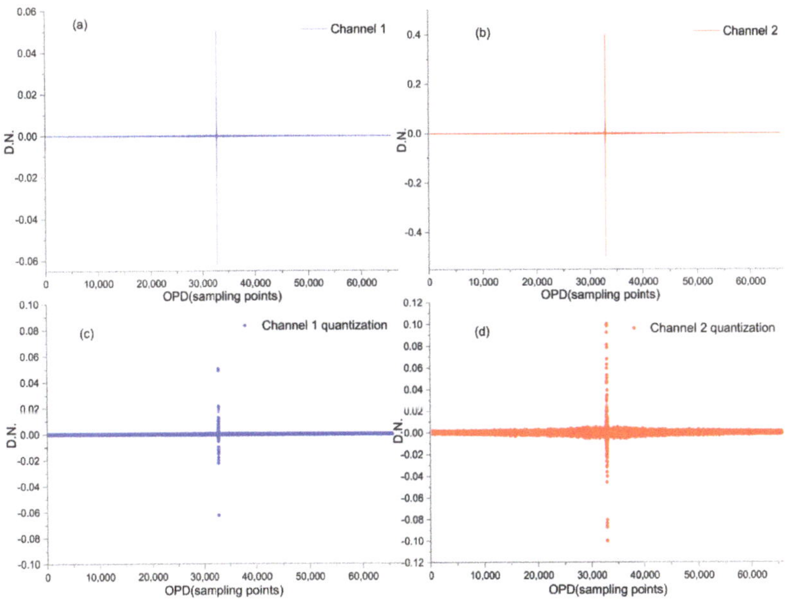

Figure 4. Dual-channel acquisition of interferograms: (**a**) interferograms at the input of ADC channel 1; (**b**) interferograms at the input of ADC channel 2; (**c**) quantization of interferograms by ADC channel 1; (**d**) quantization of interferograms by ADC channel 2.

The zero-delay coefficient, g, between intervals of gain switching is given by $g = 2^N$, where N is an integer; thus, g = 2, 4, 8, etc. All data points on the calibrated two-channel interferogram must be divided by their corresponding gain amplification to convert the entire interferogram to a uniform effective gain, prior to performing the Fourier transform (as shown in Figure 5).

After recovering the interferograms, we used the standard interferogram data processing and spectral conversion process to successfully obtain the corrected spectral data (shown in Figure 6a). In order to analyze the effect of the solar tracking system on the quality of the FTIR spectra, the performance of the infrared instrument is usually evaluated using the instrumental R_{SNR}, expressed as Equation (1), and the instrumental signal-to-noise ratio analysis is chosen to be from 6000 cm^{-1} to 6200 cm^{-1} [15]; the R_{SNR} of the reconstructed spectra of the original single-channel acquisition and the dual-channel acquisition are calculated to be, respectively, 2.128×10^3:1 and 2.613×10^3:1, and the R_{SNR} was improved by a factor of 1.23 (shown in Figure 6b).

$$R_{SNR} = \frac{100}{N} \tag{1}$$

where N is the peak noise value measured by the transmittance representation.

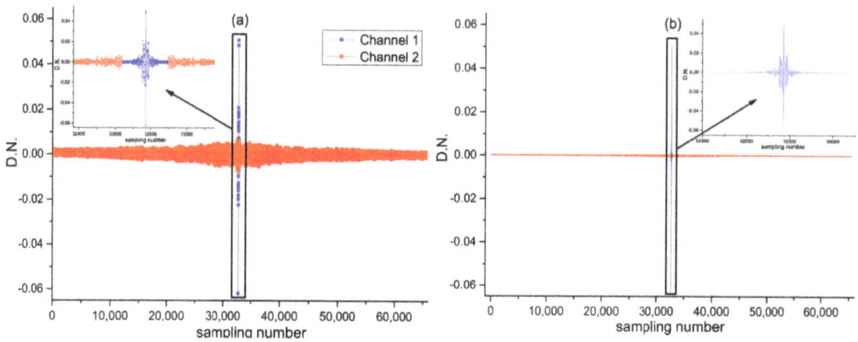

Figure 5. Dual-channel acquisition of interferogram recovery results: (**a**) acquisition of two channels of raw data; (**b**) the recovered interferogram.

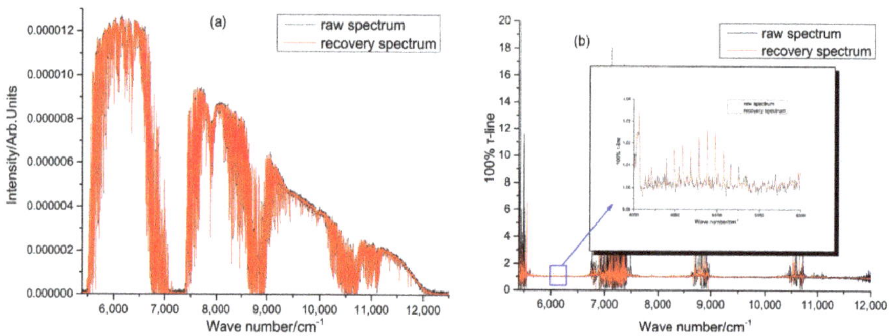

Figure 6. Analysis of the recovered spectral results: (**a**) The spectral profiles obtained by the two methods are consistent; (**b**) 100% of the τ-lines falling within the spectral range of 6000 to approximately 6200 cm^{-1}.

3.2. Hardware Design and Analysis

Matlab simulation experiments verify the feasibility of the proposed dual-channel interferometric data acquisition method, and demonstrate an improvement in spectral data quality. This section will focus on the design and analysis of several key circuit modules essential for the hardware implementation of the method, including the power supply circuit, the amplification and detection circuit for the detector signal, and the dual-channel acquisition and processing circuit.

(1) Low-noise power supply circuit design

The design of the power supply plays a crucial role in operational amplification circuits. It can directly affect the stability, performance of the circuit, and the adaptability of the application scenario. In circuit design, most often, regulated power supplies are used. The function of a regulated power supply is to obtain a stable direct current (DC) voltage from a slightly unstable power source. The main technical requirements for a regulated power supply are accuracy and stability. The power supply is also usually required to have an appropriate current output capability (preferably less than milliamps, depending on the application). There are usually two types of regulated power supplies: linear regulated-power supplies and switching regulated-power supplies, which tend to suffer from more serious noise disturbances. Therefore, we choose the purer linear regulated-power supply.

The design circuit is shown in Figure 7, with a ±15 V power supply capacity. The power supply output uses the LT1963 and LT1175 voltage regulator modules from analog devices [16], which feature very low operating current, rapid transient response, and low output noise characteristics.

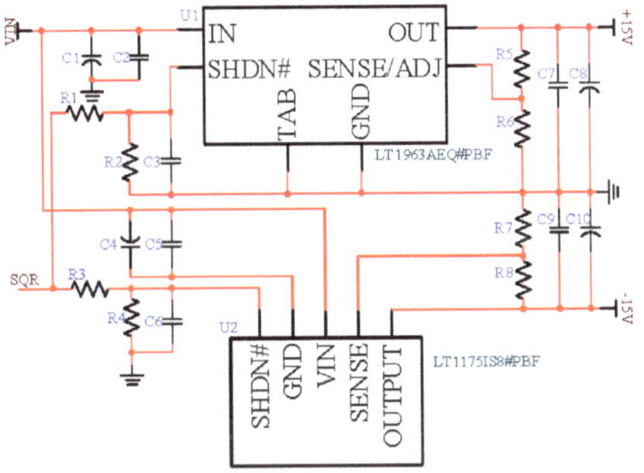

Figure 7. Power supply-circuit design schematic.

(2) Signal-detection and amplification circuit design

The solar-spectral ground-based FTIR system utilizes a high-performance InGaAs detector that operates within the spectral range of 0.7 μm to 1.8 μm. It is characterized by a fast rise time, stable temperature response, wide dynamic range, and high sensitivity. To enhance the performance, particularly the temperature stability of the response near the cutoff wavelength, it is necessary to utilize it in conjunction with a thermoelectrically cooled temperature controller. The equivalent circuit of an InGaAs photodiode, depicted in Figure 8, represents a photon-generating current source accompanied by a shunt resistor (R_D), a shunt capacitor (C_D), and a series resistor (R_S). The value of R_S is negligible compared to R_D, except at high power levels (exceeding 10 mW). The InGaAs photodiode generates current across the PN junction when photons with sufficient energy are absorbed in its active region.

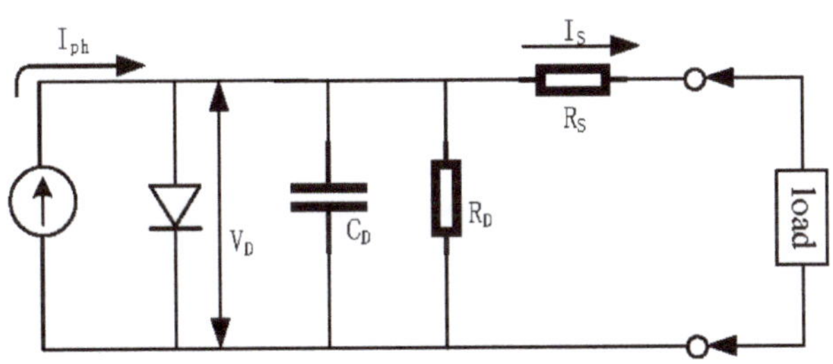

Figure 8. InGaAs photodiode equivalent circuit.

The infrared signal originally detected by the detector is relatively weak, and the design of the amplification and detection circuit must meet the technical specifications for low-noise and high-gain amplification. In Figure 7, the first stage of amplification is a negative feedback operational amplifier (op-amp). The feedback circuit converts the detector's output current to a voltage, while the op-amp maintains the detector at a near-zero-volt bias to minimize noise. However, there is a DC offset current, also known as the "dark current" of the detector circuit. This dark current is a function of the preamplifier's input bias current (I_b), the preamplifier's input offset voltage (V_{os}), and the detector's shunt resistor (R_D). The total "dark current" can be expressed by Equation (2).

$$I_D = I_b + \frac{V_{os}}{R_D} \tag{2}$$

Therefore, choosing the right operational amplifier is crucial for reducing preamplifier noise and achieving optimal system bandwidth. For a high R_D detector, select a preamplifier with a low bias current; for a low R_D detector, choose one with a low output offset voltage.

Here, we utilize the officially recommended Quest Components model OPA111, a high-precision, low-noise differential amplifier (with a maximum noise of 8 nV/$\sqrt{\text{Hz}}$ at 10 kHz), featuring low drift (with a maximum temperature drift of 1 µV/°C) and a high common-mode rejection ratio of 100 dB. The maximum input bias current is 1 pA, and the minimum open-loop gain is 120 dB.

To reduce circuit noise while maintaining the signal's gain bandwidth, the first-stage op-amp circuit must satisfy the conditions C1 = C2 and R2 × C2 = R4 × C7. The second-stage circuit provides voltage amplification and is capable of DC-coupled input and output for the IR signal (Figure 9).

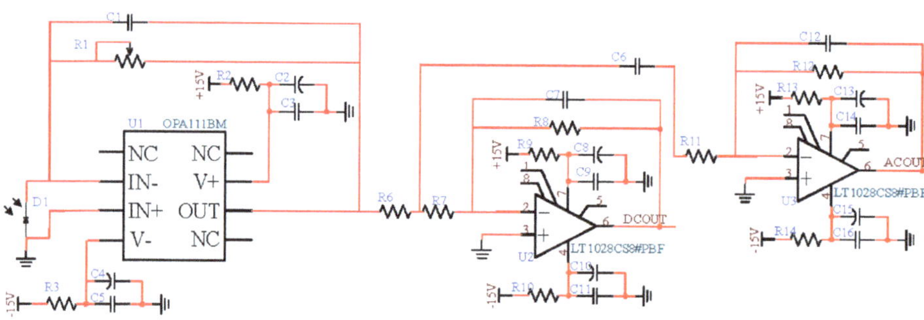

Figure 9. Interference signal amplification-detection circuit schematic.

(3) Dual-channel acquisition and processing circuit design

Based on the proposed dual-channel interferometric data acquisition method, the gain adjustment of the detection signal is realized, building upon the previous design of the power supply circuit and the signal detection and amplification circuit. Specifically, the detection signal is divided into multiple channels, each of which is amplified with a different gain. These channels are parallel to, and independent of, each other. The advantage of this approach is that each channel operates in parallel, eliminating the need for synchronization control.

Following the aforementioned infrared interferogram segmentation and gain design, the interferogram is split into two parts with different gains, necessitating two channels for amplification and collection. This setup avoids the complexity of precise synchronization control.

In response to the requirements of the circuit design, we have designed the two-stage synchronization circuit shown in Figure 10. The variation in gain is primarily determined by R3 and R9. To enhance the stability and reliability of the circuit, we have incorporated a T-capacitor filter to fulfill the filtering function of the method.

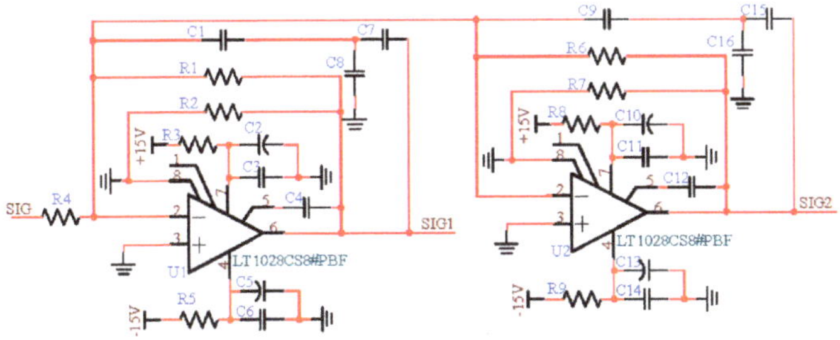

Figure 10. Schematic diagram of two-stage gain synchronous amplifier circuit.

3.3. Calibration and Correction of Channel Errors

When an ADC acquires an analog signal and converts it into a digital signal, it generates certain errors, due to various reasons such as hardware limitations, environmental factors, and circuit design imperfections. These errors may encompass nonlinearity errors, gain errors, and offset errors. Through software correction, these errors can be mitigated to a certain extent, thereby improving the acquisition accuracy of the ADC. The specific correction steps are as follows:

a. Access a known zero-voltage signal at the input of the ADC, read the output value of the ADC (which represents the zero offset of the ADC), and subtract this zero offset from subsequent ADC output values to achieve zero calibration.
b. Apply a known full-scale voltage signal to the input of the ADC, and read the ADC's output value (which corresponds to the ADC's output at full-scale voltage). The theoretical output value of the ADC at full-scale voltage should equal the reference voltage. Adjust the actual ADC output value to match the theoretical value proportionally through software programming, thereby achieving full-scale calibration.
c. Measure the ADC's output value at several known input voltage points. Based on the measured output values and the corresponding input voltage values, calculate the offset and gain factor for the ADC in each voltage range. In the software, create a look-up table to store the offsets and gain coefficients for each voltage range. During data acquisition, retrieve the corresponding offsets and gain coefficients from the look-up table based on the input voltage range, and use them to correct the ADC's output value.

4. Applications and Analysis

4.1. Experimental Instruments

In the experimental application, the self-developed AG-FTIR-GH2000 near-infrared solar absorption spectrometer was utilized to determine the column concentration of CO_2. To attain a retrieval accuracy exceeding 0.5%, the spectrometer incorporated high-precision dynamic collimated interferometry. Additionally, we utilized silver-plated optics, a calcium fluoride beamsplitter, and an InGaAs detector to achieve a spectral range spanning from 5400 to 12,800 cm^{-1}, with a resolution of 0.5 cm^{-1} and a scanning speed of 100 kHz for solar absorption spectra. The sun tracker, which integrates both an image sensor and a position sensing detector (PSD), directs sunlight into the spectrometer's entrance. The image sensor captures deviations in the sun's exact position at a rate of 3 Hz, while the position detector automatically searches for the sun's position at a rate of 100 Hz whenever the image is lost (Figure 11).

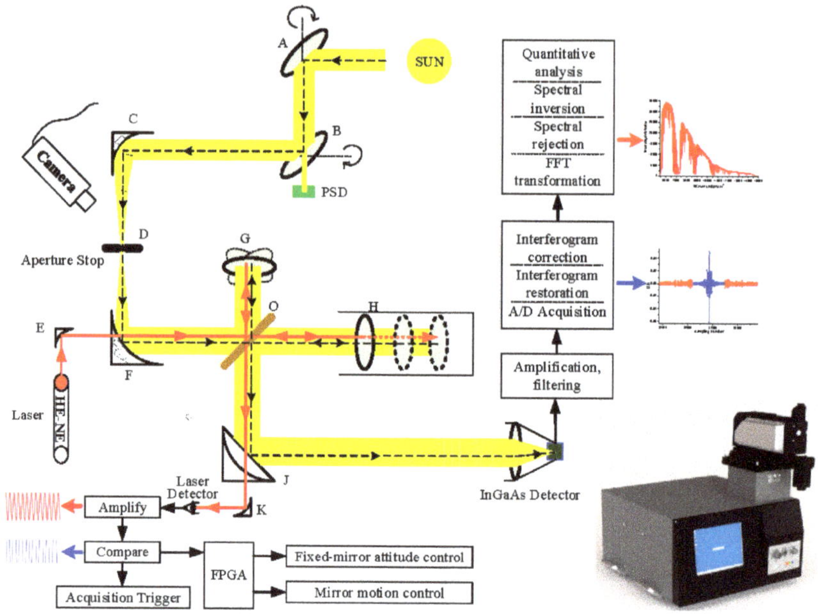

Figure 11. Solar spectrum ground-based FTIR spectrometer.

The sun tracker primarily serves as a light-source introduction system for the FTIR spectrometer. Sunlight traverses reflector A (which adjusts according to the sun's altitude angle) and reflector B (which adjusts according to the sun's azimuth angle). Reflector B features a 5 mm diameter through-hole at its center, allowing sunlight to pass through and illuminate the position detector. This detector, in turn, drives the tracking system via the controller to achieve coarse tracking of the sun's movement. The sunlight is then focused through the central field of view of a diaphragm by parabolic mirror C. Notably, the center of the diaphragm is positioned 2 cm away from the focal point. The diaphragm offers two primary advantages. Firstly, it ensures that the FTIR's resolution meets the design requirements by defining the interferometer's field of view. Secondly, any positional deviation in the beam passing through the small aperture is captured and analyzed in real-time by a camera. Based on this analysis, the controller adjusts the tracking system to achieve precise tracking of the sun's movement.

The FTIR optical system comprises the HE-NE laser interferometric path and the sunlight interferometric path. Both sets of paths share the core interferometric system, with the laser and sunlight traveling along coaxial paths. The sunlight interferometric optical path primarily consists of a parabolic mirror F, which converts the sunlight imported from the solar tracking system into a set of parallel light. This parallel light is then divided into two perpendicular beams by a beam splitter O. One of these beams is reflected upward by a fixed mirror G. The upward-reflected light is then redirected by the beam splitter O towards another fixed mirror and subsequently gathered onto the infrared detector through a reflector J, along with the other beam that passes through a moving mirror and returns to the beam splitter for final redirection. Meanwhile, the HE-NE laser interferometric path mainly consists of the laser light source, which passes through a quasi-diameter element E into the same interferometric path as the sunlight. This laser light is then gathered onto the laser detector through a reflector K. The HE-NE laser signal detection and processing system primarily involves a four-quadrant silicon PIN photodiode for detecting the interference of the laser signal. After amplification, filtering, and shaping, the processed signal is input into an FPGA for controlling the motion of moving mirrors and the attitude of fixed mirrors, as well as triggering the sampling of infrared interference signals.

The interference data acquisition and processing system primarily amplifies and filters the signals detected by the infrared detector. The interferogram is obtained through analog-to-digital (A/D) sampling and correction processes, which will constitute an important part of our testing. Spectral recovery, screening, inversion, and quantitative analysis of the interferograms are primarily conducted on a personal computer (PC).

4.2. CO_2 Observations and Analyses

In Fourier-transform spectroscopy, spectral information in the frequency domain is obtained from the time-domain output signal derived from the measurement. Therefore, once the acquisition and recovery of the interferogram are completed, the interferogram undergoes preprocessing steps such as phase correction, and subsequently, the spectrogram is obtained through FFT recovery. Figure 12 illustrates the process from the interferogram to the acquisition of the spectrogram for the ground-based FTIR spectrometer used to measure the solar spectrum.

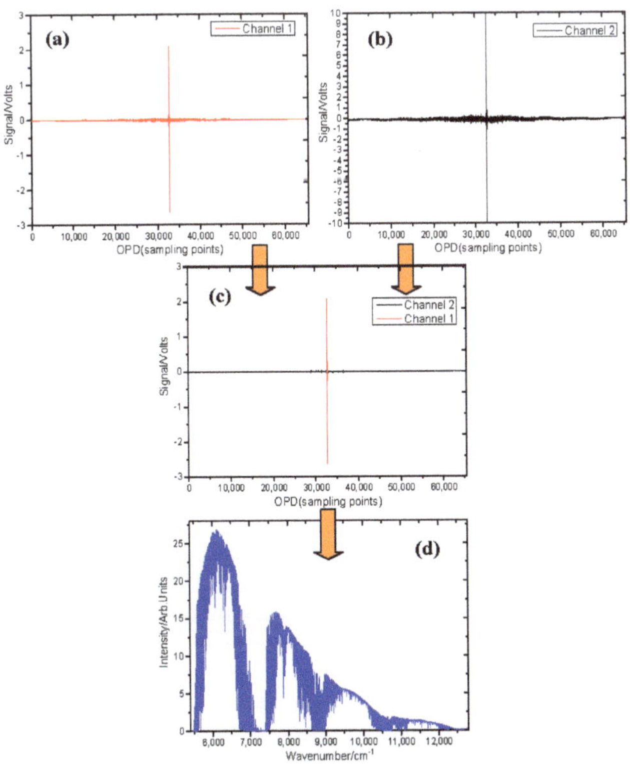

Figure 12. Interferogram acquisition to spectral recovery in field tests: (**a**) channel I; (**b**) channel II; (**c**) reconstructed interferogram; (**d**) spectrogram.

In the field of spectroscopy, Fourier-transform spectroscopy employs the time-domain output signal acquired from measurements. After undergoing mathematical transformation, we can extract the frequency-domain spectral information concealed within, thereby enabling an in-depth investigation of matter's properties. To obtain useful spectral information from the interferogram, a series of preprocessing steps must first be carried out, of which phase correction is a vital component. During the actual measurement process, various factors (such as instrumental errors, environmental interference, etc.) may cause shifts in the interferogram's phase. If left uncorrected, these shifts will directly compromise the accuracy of the final spectral map. Therefore, adjusting the interferogram's phase using precise mathematical algorithms to align with the actual physical process is crucial for ensuring the quality of spectral analysis.

To eliminate this phenomenon, the phase can be corrected using a correction term. In Fourier-transform infrared spectroscopy theory [15], the interferogram can be expressed as:

$$I(\delta) = \int_0^{\infty} I_0(v)\cdot cos2\pi v\delta\, dv \tag{3}$$

Since $\varnothing\left(\tilde{v}\right) = -\varnothing\left(-\tilde{v}\right)$, Equation (3) can be expressed as:

$$B'\left(\tilde{v}\right) = \int_{-\infty}^{\infty} I'(v)e^{-i2\pi\tilde{v}x+i\varnothing(\tilde{v})}dx = B_{real}\left(\tilde{v}\right) + iB_{imag}\left(\tilde{v}\right) \tag{4}$$

where $B_{real}\left(\tilde{v}\right)$ and $B_{imag}\left(\tilde{v}\right)$ are the real and imaginary parts.

The correction term $\varnothing\left(\tilde{v}\right)$ can be calculated by having Equation (4):

$$\varnothing\left(\tilde{v}\right) = \tan^{-1}\left(\frac{B_{imag}\left(\tilde{v}\right)}{B_{real}\left(\tilde{v}\right)}\right) \tag{5}$$

Combining Equations (4) and (5) above, an expression for the calibration spectrum can be obtained:

$$B'\left(\tilde{v}\right) = B\left(\tilde{v}\right)e^{i\varnothing(\tilde{v})} \tag{6}$$

After phase correction and other preprocessing steps, the next stage involves using the fast Fourier transform (FFT) to obtain the spectral map. The FFT can efficiently transform a vast number of data in a short time, significantly enhancing the efficiency and accuracy of spectral analysis. In this process, every detail in the interferogram is meticulously analyzed and transformed into clearly discernible spectral lines on the spectral map, each corresponding to the characteristic absorption or emission of a specific component or structure within the material.

Figure 12 depicts the transition from an interferogram to a spectrogram. A raw interferogram, magnified by a factor of 1, is utilized as the data input for ADC channel one (Figure 12a). Similarly, another raw interferogram, magnified by a factor of 8, serves as the data input for ADC channel two (Figure 12b). After dividing all the data points on the combined two-channel interferogram by their respective gain magnifications, the recovered interferogram is created by selecting the 28/2 data points around the zero-phase difference (ZPD) of channel 1's interferogram and incorporating all data from channel 2, excluding the overlapping data points from channel 1. This composite forms the final recovered interferogram (Figure 12c). It is evident that the dual-channel ADC acquires and processes the interferogram to produce a complete version, which undergoes preprocessing and FFT recovery before being converted into a spectrogram (Figure 12d).

We tested the algorithm by comparing the concentrations of CO_2 columns derived from spectra. These spectra were acquired from interferograms: some using a single channel and others using two channels. The test weather conditions were clear and cloudless. Since there are no perfect data that are completely unaffected by clouds or changes in light intensity, and since the measured gas column may be higher than the true atmospheric column, we selected data where the change in solar radiation intensity during a single scan was within 5%. On 9 October 2024, in Hefei, China (longitude 117.096386°, latitude

31.867071°), we retrieved CO_2 column concentrations from FTIR spectra measured on the roof of Science Island.

Due to Science Island's location in the center of the lake, we anticipated minimal diurnal variation in CO_2. A nearly constant CO_2 column concentration was obtained from the measurement data using a two-channel interferometric data acquisition method. This contrasts with the control data, where the dispersion of the acquired CO_2 column concentration increases when the interferometric data are acquired using the conventional method. According to the results presented in Figure 13, for the complete dataset of 487 control measurements taken between 9:24:37 and 15:28:24 on the same day, the mean CO_2 column concentration of the data acquired using the conventional method was 426.1 ppm, with fluctuations ranging from 422.0 to 430.9 ppm. The mean CO_2 column concentration of the data acquired using the dual-channel interferometric method was 426.2 ppm, with fluctuations ranging from 424.4 to 427.8 ppm.

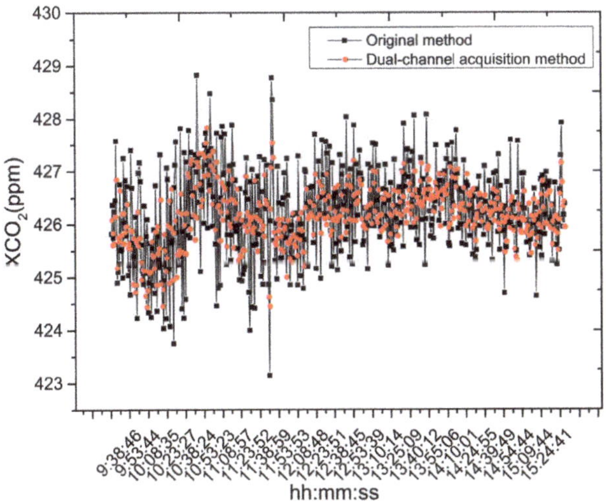

Figure 13. CO_2 column-concentration observation results.

In order to determine whether the dual-channel interferometric data acquisition method introduces bias into the gas retrieval process, we analyzed the degree of data deviation depicted in Figure 13. Across 487 measurements, the mean difference in CO_2 column concentration for data acquired using the traditional method was 0.9 ppm, with a standard deviation of 1.1 ppm. In contrast, the mean difference for data acquired using the dual-channel interferometric method was 0.4 ppm, with a standard deviation of 0.5 ppm.

Since it is difficult to simulate the real ambient atmosphere in vertical column-concentration measurements, to further demonstrate the repeatability and reliability of the CO_2 column-concentration measurement results, the ratio of the standard deviation to the daily average was employed to illustrate the variation relationship among data measured repeatedly on the same day. In 487 measurements, the accuracy of the CO_2 column concentration, as measured by the traditional method, was 0.27%. The accuracy of the CO_2 column concentration measured using the two-channel interferometric data acquisition method was 0.13%. The accuracy of the CO_2 column concentration was calculated using the following formula:

$$Gas_{precision} = \frac{\sigma}{\langle Gas \rangle} \times 100\% \tag{7}$$

Here, $Gas_{precision}$ represents the measurement precision of the gas to be measured, σ is the standard deviation of the time series, and $\langle Gas \rangle$ indicates the mean value of the measurement results.

5. Conclusions

In this paper, we propose a dual-channel ADC acquisition method for interferometric data acquisition in a ground-based Fourier-Transform Infrared (FTIR) greenhouse gas spectrometer. This method involves acquiring interferometric data through two ADC acquisition channels with different gains: the low-gain ADC channel primarily captures data near the zero optical-path difference (ZPD) spike, while the high-gain ADC channel focuses on acquiring small signals on the two flanks of the interferogram. Simulation verification results indicate that the signal-to-noise ratio is improved by a factor of 1.23. Additionally, the retrieval accuracy of CO_2 column concentration data is enhanced by a factor of 2.096, based on observations at a fixed point. We anticipate that this interferometric data acquisition method will have an increasingly significant impact on the selection of data acquisition techniques for FTIR spectrometers, particularly when the need to improve the accuracy of weak signals on the flanks of interferograms becomes more prominent, especially with limited ADC accuracy and dynamic range. In the future, we plan to continue applying this method to interferogram acquisition using a 24-bit synchronous ADC to further enhance instrument performance.

Author Contributions: Conceptualization, Y.D. and L.X.; formal analysis, Y.D. and L.X.; funding acquisition, L.J. and Y.S.; investigation, L.X.; methodology, Y.D.; resources, J.L. and W.L.; software, Y.D., L.Z., Y.S. and J.L.; validation, L.Z.; writing—original draft, Y.D.; writing—review and editing, Y.D. and L.X. All authors have read and agreed to the published version of the manuscript.

Funding: Project supported by the key R&D program of Anhui Province, grant number 2022m07020009, the National Natural Science Foundation of China, grant number 52027804 and the Special Funds of the National Natural Science Foundation of China, grant number 41941011.

Institutional Review Board Statement: Not applicable.

Informed Consent Statement: Not applicable.

Data Availability Statement: The data underlying the results presented in this paper are not publicly available at this time, but may be obtained from the authors upon reasonable request.

Conflicts of Interest: The authors declare no conflicts of interest.

References

1. Löw, B.A.; Kleinschek, R.; Enders, V.; Sander, S.P.; Pongetti, T.J.; Schmitt, T.D.; Hase, F.; Kostinek, J.; Butz, A. A portable reflected-sunlight spectrometer for CO_2 and CH_4. *Atmos. Meas. Tech.* **2023**, *16*, 5125–5144. [CrossRef]
2. Humpage, N.; Boesch, H.; Okello, W.; Chen, J.; Dietrich, F.; Lunt, M.F.; Feng, L.; Palmer, P.I.; Hase, F. Greenhouse gas column observations from a portable spectrometer in Uganda. *Atmos. Meas. Tech. Discuss.* **2023**, *2023*, 5679–5707. [CrossRef]
3. Mostafavi Pak, N.; Hedelius, J.K.; Roche, S.; Cunningham, L.; Baier, B.; Sweeney, C.; Roehl, C.; Laughner, J.; Toon, G.; Wennberg P.; et al. Using portable low-resolution spectrometers to evaluate Total Carbon Column Observing Network (TCCON) biases in North America. *Atmos. Meas. Tech.* **2023**, *16*, 1239–1261. [CrossRef]
4. Davis, S.P.; Abrams, M.C.; Brault, J.W. *Fourier Transform Spectrometry*; Academic Press: Cambridge, MA, USA, 2001.
5. Duan, M.; Zong, Y.; Zhu, R.; Li, J. Phase-tilt iteration: Accurate and robust phase extraction from random tilt-shift interferograms. *Opt. Lasers Eng.* **2021**, *142*, 106595. [CrossRef]
6. Shu, S.; Liu, J.; Xu, L.; Wang, Y.; Deng, Y.; Sun, Y. Real-Time Simulation of Clear Sky Background Radiation in Gas Infrared Remote Sensing Monitoring. *Photonics* **2024**, *11*, 904. [CrossRef]
7. Palchetti, L.; Lastrucci, D. Spectral noise due to sampling errors in Fourier-transform spectroscopy. *Appl. Opt.* **2001**, *40*, 3235–3243 [CrossRef]

8. Griffiths, P.R.; de Haseth, J.A. *Fourier Transform Infrared Spectrometry*, 2nd ed.; Analytical and Bioanalytical Chemistry; Springer: Berlin/Heidelberg, Germany, 2008; Volume 391, pp. 2379–2380.
9. Cohen, D.L. Performance degradation of a Michelson interferometer due to random sampling errors. *Appl. Opt.* **1999**, *38*, 139–151. [CrossRef] [PubMed]
10. Simon, A.; Metz, W.; Keens, A. Data acquisition and interferogram data treatment in FT-IR spectrometers. *Vib. Spectrosc.* **2002**, *29*, 97–101. [CrossRef]
11. Alber, G.M.; Marshall, A.G. Effect of sampling rate on Fourier transform spectra: Oversampling is overrated. *Appl. Spectrosc.* **1990**, *44*, 1111–1116. [CrossRef]
12. Brault, J.W. New approach to high-precision Fourier transform spectrometer design. *Appl. Opt.* **1996**, *35*, 2891–2896. [CrossRef] [PubMed]
13. Kang, Y. Analysis of sampling error in FTIR. In Proceedings of the PIAGENG 2013: Image Processing and Photonics for Agricultural Engineering, Sanya, China, 27–28 January 2013; SPIE: Bellingham, WA, USA, 2013; Volume 8761, pp. 122–127.
14. Hao, J.; Lu, C.; Jiang, W. Research on interference signal acquisition technology of Fourier transform infrared spectrometer. *Autom. Technol. Appl.* **2016**, *10*, 23–27.
15. Griffiths, P.R. Fourier transform infrared spectrometry. *Science* **1983**, *222*, 297–302. [CrossRef] [PubMed]
16. Kester, W. (Ed.) *Practical Design Techniques for Power and Thermal Management*; Analog Devices: Wilmington, MA, USA, 1998.

photonics

MDPI

Article

Dual-Stream Enhanced Deep Network for Transmission Near-Infrared Dorsal Hand Vein Age Estimation with Attention Mechanisms

Zhenghua Shu, Zhihua Xie * and Xiaowei Zou

Key Lab of Advanced Electronic Materials, Jiangxi Science and Technology Normal University, Nanchang 330031, China; 1020101066@jxstnu.edu.cn (Z.S.); 1020101062@jxstnu.edu.cn (X.Z.)
* Correspondence: xie_zhihua68@aliyun.com; Tel.: +86-139-7001-8047

Abstract: Dorsal hand vein recognition, with unique stable and reliable advantages, has attracted considerable attention from numerous researchers. In this case, the dorsal hand vein images captured by the means of transmission infrared imaging are clearer than those collected by other infrared methods, enabling it to be more suitable for the biometric applications. However, less attention is paid to individual age estimation based on dorsal hand veins. To this end, this paper proposes an efficient dorsal hand vein age estimation model using a deep neural network with attention mechanisms. Specifically, a convolutional neural network (CNN) is developed to extract the expressive features for age estimation. Simultaneously, another deep residual network is leveraged to strengthen the representation ability on subtle dorsal vein textures. Moreover, variable activation functions and pooling layers are integrated into the respective streams to enhance the nonlinearity modeling of the dual-stream model. Finally, a dynamic attention mechanism module is embedded into the dual-stream network to achieve multi-modal collaborative enhancement, guiding the model to concentrate on salient age-specific features. To evaluate the performance of dorsal hand vein age estimation, this work collects dorsal hand vein images using the transmission near-infrared spectrum from 300 individuals across various age groups. The experimental results show that the dual-stream enhanced network with the attention mechanism significantly improves the accuracy of dorsal hand vein age estimation in comparison with other deep learning approaches, indicating the potential of using near-infrared dorsal hand vein imaging and deep learning technology for efficient human age estimation.

Keywords: transmission near-infrared; attention mechanism; dorsal hand vein; age estimation; the dual-stream enhanced network with attention mechanism (DSEANET)

1. Introduction

Automatic age estimation, which estimates individual ages based on facial images, is a highly regarded topic in pattern recognition, machine learning, and computer vision [1]. In 1994, Kwon et al. published early research on age classification theory and practical methods for facial images based on craniofacial features and skin wrinkle analysis [1]. By 2002, Lantis et al. [2] used learning-based age transformation features to explain the impact of age on facial appearance, indicating that reliable and accurate age estimation can be performed on facial images. Subsequently, with the establishment of numerous age recognition-related datasets, researchers have conducted more extensive and in-depth research on age estimation [3].

Early physiological age estimation methods used traditional machine learning algorithms and can be generally divided into two stages. The first stage involved manually extracting age-related features, and the second procedure employed the extracted features to optimize the age estimation model through various algorithms to accomplish classification and estimation tasks. With the advancement of deep learning technology, some works have gradually applied deep learning to facial age estimation, achieving encouraging performance. Especially when the training samples are large enough, the performance of convolution neural network (CNN) models becomes more excellent. It can learn diverse and compact global features, thus strengthening the feature representation ability of the models. Therefore, it has been successfully applied to face analysis tasks, such as face detection, face recognition [4], face verification [5], and demographic estimation [6]. Following this principle, Niu et al. [7] developed an end-to-end facial age estimation model based on CNNs. Sendik et al. [8] proposed a facial age estimation approach in combination with CNN and SVR, which has achieved good results on small-scale datasets. Chen et al. [9] designed an unconstrained age estimation method based on a depth convolutional neural network (DCNN). Zhang et al. [10] designed a multi-level residual network as an age estimation model. The DCNN first pre-trains the model on the Image Net dataset and then optimizes it on the IMDB-WIKI-101 dataset. Tao et al. [11] came up with a new facial age estimation method based on the deep forest model. Liu H et al. [12] presented a label-sensitive deep metric learning method for facial age estimation. Taheri S et al. [13] combined the multi-layer features in a CNN with a series of age-related manual features to estimate facial age. However, the visible facial imaging usually varies with the ambient illumination and disguise, which makes facial age estimation extremely challenging.

Fortunately, physiological tissues are responsive to near-infrared light, which typically ranges between 700 and 1000 nm in wavelength. When the hand is exposed to near-infrared light of a specific wavelength, the light penetrates the epidermis and enters the subcutaneous tissue, where it scatters. During the scattering process, the near-infrared light is significantly absorbed by deoxyhemoglobin in the venous blood, resulting in a dark shadow at the location of the hand vein when captured by an image sensor, while regions without veins appear brighter. The absorption of near-infrared light by deoxyhemoglobin at the vein sites ensures that the veins appear as dark shadows in the captured images, whereas non-venous areas remain brighter [14]. This unique property of dorsal hand vein images contributes to their high stability and robust resistance against spoofing attempts.

The distribution of subcutaneous veins varies among individuals, yet this distribution remains relatively stable over time, making venous patterns reliable physiological characteristics for personal identity authentication [14]. Based on this finding, reflective near-infrared methods are commonly used to capture dorsal hand veins. Nevertheless, the original vein images captured by this method may be very blurry. To address this issue, Shu et al. designed a transmissive imaging system to more clearly capture the dorsal hand vein images. They further trained and applied deep residual networks with attention mechanisms for effective feature representation and recognition of dorsal hand vein images [15].

Despite the significant advancements in biometric research based on dorsal hand vein images, most scholars have concentrated on identity recognition, with relatively little attention given to individual age estimation. However, it is statistically significant that dorsal hand vein images reveal changes as individuals age. Specifically, younger individuals tend to have thinner dorsal hand veins, which become thicker and more pronounced with age. Meanwhile, as people age, their veins become more prominent and curved. More importantly, unlike the facial images, the dorsal hand veins remain relatively constant and cannot be easily disguised. In a word, near-infrared dorsal hand vein imagery presents a promising solution for age estimation.

Currently, the deep learning networks are widely applied in biometric fields and have shown good performance. For instance, He et al. [16] proposed a deep residual network construction method based on the residual learning algorithm. Zhu et al. introduced a convolutional attention module [17], which is used for facial recognition. Lee et al. [18] designed a lightweight multitasking convolution algorithm to reduce model size. Although this algorithm has good real-time performance, its accuracy is not satisfactory. Lin et al. [19] developed the method of data enhancement and improved loss function to solve the problem of face age and gender prediction under occlusion, but the network structure of this algorithm is extremely complex. Wang et al. came up with a facial age estimation algorithm with a dual-stream enhanced shallow network under unrestricted conditions [20]. Given the strong learning capability of deep models, it is worthwhile to explore suitable models for dorsal hand vein age estimation. Currently, according to the public literature, most methods for dorsal hand vein recognition are CNNs or ResNets [21]. Specifically, Kuzu et al. explored the similarity of vein patterns in the finger vein, palm vein, and dorsal vein of a person's left and right hands based on deep learning-based feature [22]. Cimen et al. developed an advanced and complex fractal technique for extracting features from the images of hand vessel patterns [23].

Dorsal hand vein age estimation provides an alternative solution besides facial age estimation. The main motivation of our work is to explore an efficient deep learning model for dorsal hand vein age prediction equipped with a suitable vein image capturing system. Specifically, this paper proposes a dorsal hand age estimation system based on transmission near-infrared and the dual-stream enhancement network with diverse attention mechanisms, which can effectively extract discriminant age-aware features. The main contribution of this work indicates the potential of near-infrared dorsal hand vein imaging and deep learning techniques for efficient human age estimation.

2. Collection and Characteristics of Transmission Near-Infrared Dorsal Hand Vein Images

This study constructs a transmission-based capturing system for collecting near-infrared dorsal hand vein images, as shown in Figure 1 [15].

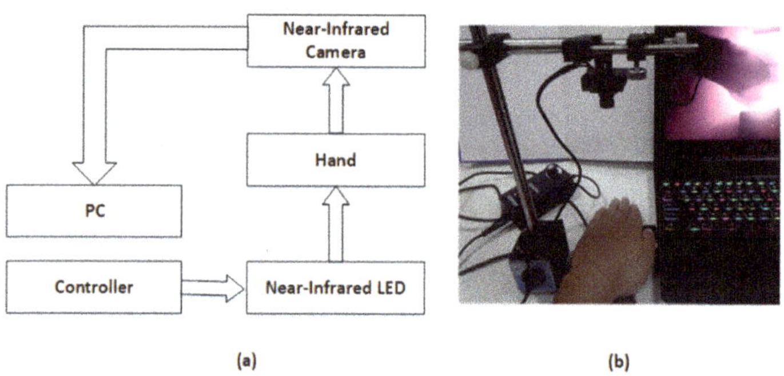

Figure 1. The schematic diagram and three-dimensional layout of near-infrared dorsal hand vein capturing system [15]. (**a**) the component diagram; and (**b**) the physical image.

Specifically, we designed a near-infrared dorsal hand vein capture system, which is composed of a light source component, a collection component, and a peripheral component. In the transmission near-infrared dorsal hand vein image collection system (Figure 1), an 850 nm LED with a 7 cm $\times$ 7 cm area light source was positioned beneath the palm. The wide motion camera (DAHUA, 3200_1080p), sensitive to wavelengths between 700 and 1100 nm, was arranged above the palms. This setup highlights the veins through near-infrared transmission, enabling the camera to capture clear dorsal hand vein images. Based on this configuration, a database of near-infrared dorsal hand vein images based on the transmission near-infrared spectrum was established for different ages.

Figure 2 provides partial near-infrared dorsal hand vein images based on transmittance across different age groups. The experiment primarily focused on adults, collecting images from 300 individuals. The age of the samples is roughly evenly distributed from 10 to 85 years old. In total, 1800 samples were obtained, with each participant providing six samples, thus forming a benchmark database of near-infrared hand vein images. As reported in reference [3], compared to the original near-infrared hand vein images obtained by reflection [15], the collected near-infrared hand vein images based on transmission are clearer. From Figure 2, it is evident that the veins on the dorsal hand change as people grow and age. In younger individuals, the dorsal hand veins are thinner. As people mature, these veins become more pronounced. With further aging, the dorsal hand veins become even more prominent and curved. Additionally, aging leads to an increase in the number of wrinkles on the skin of the dorsal hand, as shown in Figure 2s,t,u. The veins tend to become more curved with age, as illustrated in Figure 2q–u.

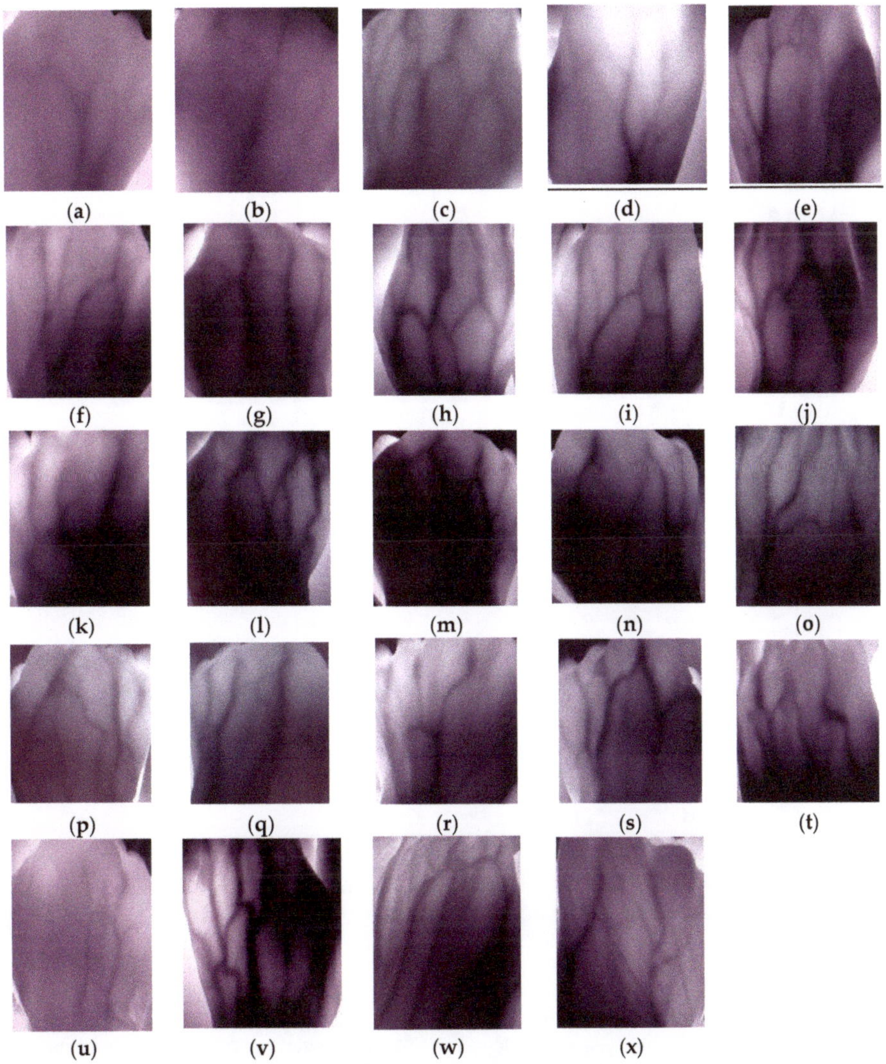

Figure 2. Transmission near-infrared dorsal hand vein at some ages: (**a**) 10 years old; (**b**) 15 years old; (**c**) 17 years old; (**d**) 20 years old; (**e**) 24 years old; (**f**) 26 years old; (**g**) 31 years old; (**h**) 35 years old; (**i**) 39 years old; (**j**) 41 years old; (**k**) 43 years old; (**l**) 48 years old; (**m**) 52 years old; (**n**) 54 years old; (**o**) 55 years old; (**p**) 56 years old; (**q**) 60 years old; (**r**) 62 years old; (**s**) 64 years old; (**t**) 66 years old; (**u**) 69 years old; (**v**) 72 years old; (**w**) 77 years old; and (**x**) 85 years old.

Figure 3 shows the Local Binary Pattern (LBP) image [24] and LBP histogram [24] of hand dorsal vein images for different ages. It can be seen that the LBP histogram of the dorsal hand veins varies with age. In younger individuals, the LBP histogram distribution of the dorsal hand veins is relatively uniform. As individuals mature, the distribution of LBP histograms becomes more uneven. However, the LBP histogram distribution tends to return to a more uniform pattern in older individuals. Therefore, it can be inferred that the statistical distribution of texture features in different ages tends to have significant discrepancies.

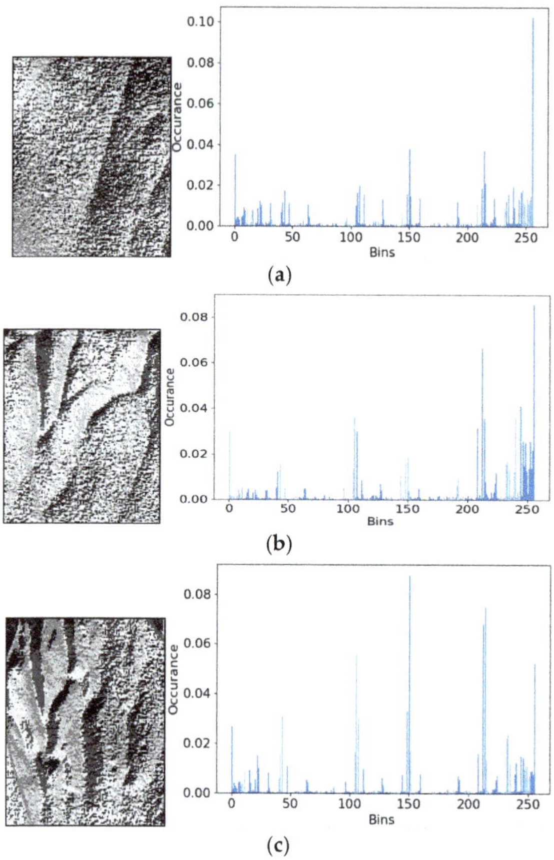

Figure 3. LBP and LBP histogram. (**a**) The LBP image (**left**) and LBP histogram (**right**) of the 15 year old; (**b**) the LBP image (**left**) and LBP histogram (**right**) of the 26 year old; and (**c**) the LBP image (**left**) and LBP histogram of the 72 year old.

3. The Proposed Approach

Deep learning networks can automatically extract high-dimensional features from near-infrared dorsal hand vein data. However, due to the extensive number of convolution and pooling operations, CNNs can also lose some important features of near-infrared dorsal hand veins, leading to an insufficient representation of the data features extracted by the network. To cope with this problem, this paper proposes a dual-stream enhancement network with an attention mechanism for dorsal hand vein age estimation. The network model comprises a CNN and a residual network. Additionally, an adaptive fusion module based on attention mechanisms is incorporated into the proposed model to achieve the dual enhancement effect. This integration operator enables the model to capture the subtle aging features of near-infrared dorsal hand veins more effectively.

3.1. The Dual-Stream Enhanced Network with Attention Mechanism

To address the issue of insufficient feature extraction ability in a single neural network model, a dual-stream enhancement network model with attention for estimating the dorsal hand veins age is proposed. The dual-stream enhancement network model is composed of a convolutional neural network embedded with an attention mechanism and a residual network with an attention mechanism. Each single network model independently extracts the features of the dorsal hand vein information and then separately predicts the dorsal hand vein age. For the first stream, the basic structure of the convolutional neural network with attention comprises a convolution module, an attention module, batch normalization (BN), a nonlinear hyperbolic tangent function (Tanh) activation function, and the maximum pooling layer. With regards to the second stream, the structure of the residual network with attention includes a residual network, ReLU activation function, average pooling, and the attention module.

The primary difference between the two single-stream networks lies in their activation functions and pooling strategies. Due to the good performance of the Tanh in distinguishing feature differences, making full use of the nonlinearity of the Tanh function is particularly beneficial for handling the dorsal hand vein characteristics with obvious differences in datasets, such as those between children and adults. In contrast, the residual network model with attention employs the ReLU activation function, which helps each single-stream network learn distinct features, thereby achieving complementary advantages. The detail framework of the dual-stream enhanced network model with attention mechanisms is illustrated in Figure 4. Initially, the output of the first fully connected layer in two single-stream shallow convolutional neural networks is concatenated to perform feature fusion. Then, it goes through two fully connected layers containing 512 neurons. In the end, the SoftMax classifier from the fused stream is used to make the final age prediction, which outputs the probability of belonging to a certain person in the age stage label.

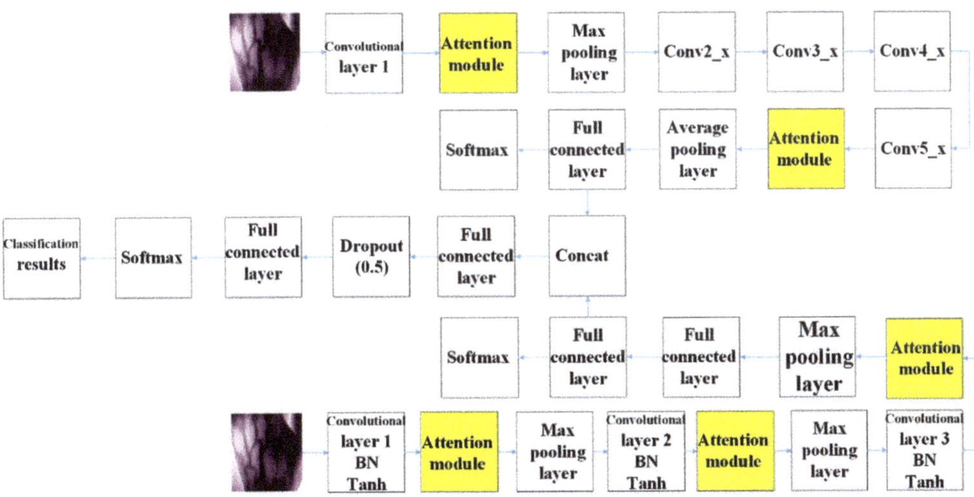

Figure 4. The framework of the dual-stream enhanced network with an attention mechanism (DSEANET).

3.2. Dual Enhanced Convolutional Attention Module

The attention mechanism focuses on the relative relationship between the target area and the relationship of key features in the image, making it very suitable for images containing structured objects, such as the dorsal hand veins. Based on the above characteristics and to deal with the problem of feature information loss during extraction, the convolutional attention module (CAM) proposed by Zhu et al. [17] was introduced to the dual-stream enhanced network, forming a dual enhancement effect. This integration enables the model to capture subtle aging features of the dorsal hand vein. The CAM module includes two parts, namely, the channel attention module and the spatial attention module. Considering the given feature map, attention weights are derived along the two dimensions of channel and space. These weights are then multiplied with the original feature map to dynamically adjust the balance between useful and irrelevant features. This process enables the network to learn global features and enhances the model's generalization ability. The main intention of the channel attention module is to highlight useful hand vein features while suppressing irrelevant information. In contrast, the spatial attention module focuses on identifying significant features of the dorsal hand veins and tracking changes in the hand vein contours.

3.3. Dorsal Hand Vein Age Estimation Based on the Enhanced Dual-Stream Network with Attention Mechanism

The DSEANET incorporates an attention mechanism to accelerate the network's feature extraction capability via the feature fusion module. The DSEANET employs diverse pooling and activation functions to address the challenge of easily losing certain features. To further facilitate the network in learning global features and balancing the confidence level of feature maps across various spatial and channel dimensions, we integrate the CAM (Class Activation Map) module into the dual-stream enhanced network model. This integration confirms the advantageousness of the DSEANET in furthering the identification of hand vein images in age estimation models. It enhances the model's nonlinearity and feature recalibration, thereby enhancing the network's classification ability and robustness.

A perfect neural network model not only excels on the training dataset but also exhibits robust performance on the testing set. To minimize errors on the test set, meet real-time requirements, and enhance the overall generalization ability, regularization of the network model is imperative. Unlike single-stream network models, the dual-stream enhanced network model expands network parameters due to an increased network width, thus elevating the risk of overfitting, particularly with small-scale data. To capitalize on the advantages of the dual-stream enhanced network model while mitigating potential drawbacks, it is crucial to employ suitable strategies to prevent overfitting. Given the basic network model's three-layer architecture, the fully connected layer of the DSEANET utilizes the Dropout (0.5) method to counter overfitting.

When estimating the age of an unknown dorsal hand vein image, the process involves several steps. Initially, a single-stream CNN network and a residual network are employed to simultaneously extract features from the dorsal hand veins. Subsequently, fundamental aging characteristics are derived through a sequence of convolutional, pooling, and batch normalization operations. Following this, a convolutional attention module is utilized to further extract and filter aging features from the dorsal hand veins, considering both channel and spatial dimensions. This ensures that pertinent features receive greater emphasis while disregarding irrelevant or less significant ones. Finally, the extracted features are flattened and fed into two pathways: in the first pathway, they are inputted into the last two fully connected layers of each single-stream network for age prediction, while in the second pathway, they are fused with features extracted from the two single-stream networks and then inputted into two fully connected layers for age estimation.

4. Experiments and Results

4.1. Experimental Platform and Dataset

The experimental setup utilized an Ubuntu 19.10 running on the TensorFlow framework, implemented in Python. The hardware configuration facilitated the efficient execution of the deep learning tasks, particularly the computationally intensive operations involved in training and inference processes.

The experiment utilized a database comprising dorsal hand vein images captured by a near-infrared transmission acquisition instrument. This database contains dorsal hand vein images from 300 individuals, with 3 images of the left hand and 3 images of the right hand captured for each person, resulting in a total of 1800 images. The age of the samples is roughly evenly distributed from 10 to 80 years old: 41 people for 10~20 years old, 42 people for 21~30 years old, 43 people for 30~40 years old, 45 people for 40~50 years old, 42 people for 50~60 years old, 45 people for 60~70 years old, and 42 people for 70+ years old. An example of the dorsal hand vein images can be seen in Figure 2. For the experiment, 800 dorsal hand vein images of the left or right hands from 200 individuals were allocated for the training set, while the remaining 1000 hand vein images captured using near infrared were designated for the test set.

All the input images were uniformly processed and resized to 227 × 227 before being fed into the network model. During the training process, each iteration consisted of 800 images. To further optimize the network model, the random gradient descent algorithm, cross-entropy loss function, and Momentum optimizer were employed for training. The initial learning rate was set to 0.01, with a maximum of 2000 iterations. The momentum was fixed at 0.9. The decay factor was set to 0.95, the decay steps to 80, and a decay coefficient of 0.1 was applied. The batch size during training was set to 80. After 200 epochs of training iterations, a network model for classification was obtained. The training process of the DSEANET, the deep residual network with attention mechanism (DRNAM), and ResNet network models is illustrated in Figure 5.

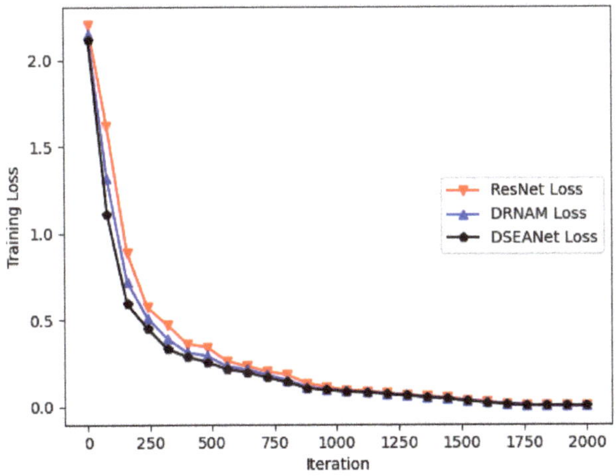

Figure 5. The training loss curves of different models.

This visualization provides insights into the progression of the training process and the convergence of the models over iterations. Figure 5 clearly illustrates that during the initial training period, although both the DSEANET and ResNet exhibit similar convergence trends, the DSEANET method demonstrates a favorable advantage in achieving lower loss values. Specifically, the loss value of the DSEANET method after training is notably smaller compared to both ResNet and the DRNAM. This observation underscores the effectiveness and rationality of the DSEANET algorithm in optimizing the model and enhancing its performance. By prioritizing smaller loss values early in the training process, the DSEANET showcases its capability to efficiently learn and extract relevant features from the input data, thus contributing to improved model convergence and overall performance.

Figure 6 reveals significant fluctuations in the recognition rate of the ResNet algorithm with changes in iterations. Conversely, the DSEANET algorithm consistently demonstrates superior estimation performance on the dataset of dorsal hand vein images captured using transmission near infrared. Particularly noteworthy is that during the early training iterations, the estimation performance of the DSEANET algorithm surpasses that of ResNet. Furthermore, after stabilizing around 200 training iterations, the recognition rate of the

DSEANET remains consistently higher than that of ResNet as the training period progresses. This trend indicates the robustness and effectiveness of the DSEANET algorithm in achieving superior recognition performance on dorsal hand vein images, especially in comparison to ResNet, across various stages of the training process.

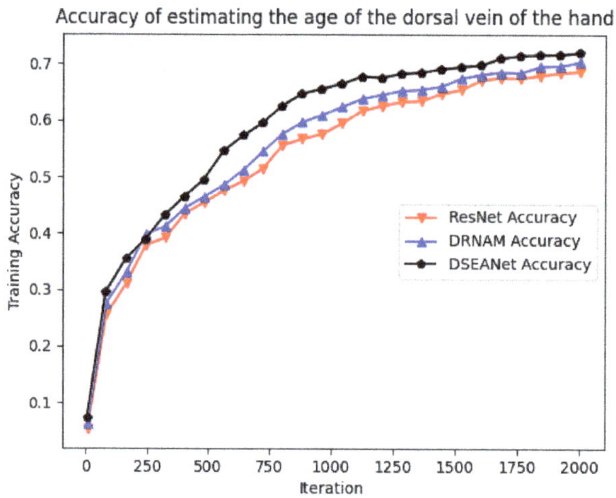

Figure 6. Training accuracy of different models.

4.2. Experimental Results and Analysis

The experiments utilized a database comprising near-infrared dorsal hand vein images captured using the transmission near-infrared spectrum to investigate and compare the age estimation capabilities of different algorithms.

To evaluate the effectiveness of the DSEANET model in estimating the age of individuals across different age groups, samples ranging from 10 to 70 years old were segmented into intervals of 10 years each. Consequently, the final dataset of near-infrared dorsal hand vein images was divided into 7 age groups: 10–20, 21–30, 31–40, 41–50, 51–60, 61–70, and 70+. The estimated results obtained from the DSEANET model, ResNet, and the deep residual network with attention mechanism (DRNAM) model for the seven age groups in the dorsal hand vein images based on the reflection-type near-infrared spectrum are presented in Table 1.

Table 1. Accuracy (%) of estimating the age of the dorsal vein of the hand on different CNNs.

Age Groups	Neural Network Models		
	DSEANET	DRNAM	ResNet
10~20 years old	83	81	79
21~30 years old	70	68	66
30~40 years old	65	63	61
40~50 years old	63	61	59
50~60 years old	72	70	69
60~70 years old	71	69	68
70+ years old	82	81	79
Average accuracy	72.0	70.4	68.7

From Table 1, it is obvious that the age estimation accuracy of the DSEANET is higher than that of the other two deep networks on the near-infrared dorsal hand vein image database. The integration of the attention mechanism empowers the DSEANET to effectively emphasize salient age characteristics within near-infrared dorsal hand vein images, thus boosting the accuracy of age estimation. Upon further comparison of the results in Table 1, it is observed that the overall accuracy of age estimation for middle-aged individuals in the dorsal hand vein dataset based on reflection-type near infrared has decreased. This decline is attributed to the gradual and subtle changes in age characteristics of dorsal hand veins throughout adulthood (21–50 years old). The minimal variations in age characteristics among different adult age groups hinder effective feature learning and subsequently impact the accuracy of age estimation.

To verify the effectiveness of the DSEANET algorithm, it was compared with other age estimation algorithms on the near-infrared hand vein dataset. The precision and recall metrics for each algorithm are presented in Table 2.

Table 2. Age classification results of hand vein dataset.

Neural Network Model	Precision (%)	Recall (%)
ResNet	59.5	82.4
DRNAM	61.2	83.6
DSEANET	62.7	77.3

As shown in Table 2, the precision of the DSEANET is better than that of ResNet and the DRNAM, indicating a higher accuracy in correctly identifying age groups. However, the recall of the DSEANET is smaller compared to ResNet and the DRNAM, implying that it may miss some instances of certain age groups. Despite this, the overall performance of the dorsal hand vein age estimation method based on transmission near-infrared, leveraging the DSEANET, remains notably superior to other near-infrared hand vein age estimation algorithms.

The combined insights from Tables 1 and 2 provide compelling evidence of the effectiveness of the DSEANET in estimating the age of dorsal hand veins. While precision reflects the model's ability to make accurate predictions, recall highlights its capability to capture instances of each age group effectively. Although there may be slight trade-offs between precision and recall, the superior performance of the DSEAET confirms its suitability and reliability for age estimation tasks in dorsal hand vein images.

Moreover, the Mean Absolute Error (MAE) and Deviation Coefficient (ε_{error}) are two important measurement criteria to further evaluate the performance of facial age estimation [25–27]. Here, we also use the two metrics, MAE and ε_{error} [28], to present the superiority of the proposed dorsal hand vein age estimation model.

Firstly, we can calculate the difference between the predicted age and the actual age obtained by estimating the age of the dorsal hand vein in the test sample image. Then, the absolute averages of the differences between all the samples are processed to obtain the MAE. The computing formula is shown in Equation (1).

$$MAE = (1/\hat{N})\sum_{i=1}^{\hat{N}}|a_i - \hat{a}_i| \quad (1)$$

where $\hat{N}$ represents the total number of samples to be tested; a_i and $\hat{a}$ are the label age and predicted age of the i-th sample. The smaller the *MAE* value, the better the performance of the algorithm and the higher the accuracy of age estimation.

The specific formula for ε_{error} is shown in Equation (2) [28].

$$\varepsilon_{error} = 1 - (1/N)\sum_{i=1}^{N} e^{\frac{(\hat{b}_i - b_i)^2}{2\hat{\sigma}_i^2}} \quad (2)$$

where $\hat{\delta}_i$ is the standard deviation of the predicted age, $\hat{b}_i$ is the average predicted age, and b_i is the actual age. ε_{error} ranges from 0 (best) to 1 (worst); the closer it is to 0, the more accurate the estimation is.

Considering the network models, the pre-trained ResNet, DRNAM, and DSEANET are selected for further experimental comparison, with performance indicators of the MAE and ε_{error}, respectively. The experimental results on the dorsal hand vein dataset are shown in Table 3.

Table 3. Comparison of accuracy results of hand vein dataset. Bold fonts indicate the best results.

Neural Network Models	MAE (Year)	ε_{error}
ResNet	5.637	0.437
DRNAM	5.271	0.418
DSEANET	5.138	0.395

Compared with other models, the DSEANET achieved better experimental results on the near-infrared dorsal hand vein dataset. The main reason for the outstanding performance of the DSEANET is that this model adopts two heterogeneous single-stream neural networks to improve the learning ability. This is because the convolutional attention module (CAM) introduced by the DSEANET can refine shallow features into high-level features through attention modules. The CAM also enhances the network's ability to extract high-level features and improves the differentiation and expression of critical age information. At the same time, the combination of the residual structure and improved attention module can better express the correlation between different channel features, thereby enhancing the expressive power of refined attention features. On the other hand, there will be no network degradation due to network deepening, thus achieving better recognition performance. The serial structure of the attention module and convolutional layer first enhances the representation of key features in the feature map and then extracts features more in line with the human visual mechanism. The DSEANET has shown significant improvements compared to ResNet and the DRNAM in all aspects, proving the feasibility of the DSEANET.

5. Conclusions

To enhance the generalization ability of dorsal hand vein age estimation, this paper proposes a novel dual-stream enhanced network that integrates attention mechanism modules. This model, built upon the foundation of CNNs, comprises two streams: a convolutional neural network and a residual network, both augmented with attention mechanisms. By leveraging attention mechanisms, the dual-stream enhanced network model enhances the learning capacity of the network. To address the issue of feature information loss during feature extraction, the two constituent single-stream networks employ distinct activation and pooling functions. The experimental results demonstrate that the dual-stream enhanced network model with attention mechanisms achieves notable success in estimating age based on transmission near-infrared dorsal hand vein images. However, samples of our hand vein database are relatedly limited for deep neural networks and there is still room for good generalization on the gender and ethnicity factors. In the future, more training and testing samples can be captured to facilitate training a deeper model with the self-attention mechanism, which can further enhance the performance of age estimation. Meanwhile, the differences in dorsal hand vein patterns between males and females will be explored for the robust age estimation system.

Author Contributions: Conceptualization and methodology, Z.S.; software, Z.X.; validation, X.Z.; formal analysis, X.Z.; investigation and resources, Z.X.; data curation, X.Z.; writing—original draft preparation, Z.S.; writing—review and editing, Z.X.; visualization, Z.S.; supervision, Z.X.; project administration, Z.X.; funding acquisition, Z.X. All authors have read and agreed to the published version of the manuscript.

Funding: This work was funded by the National Nature Science Foundation of China (No. 62362037 and No. 12264018), the Natural Science Foundation of Jiangxi Province of China (No. 20224ACB202011), and the Jiangxi Province Graduate Innovation Special Fund Project (No. YC2022-s790).

Institutional Review Board Statement: Our studies were conducted under IRB of Jiangxi Science and Technology Normal University approval, and the following is the information. Approval Code: IRB-JXSTNU-2023007, Approval Date: 2 April 2023.

Informed Consent Statement: Informed consent was obtained from all subjects involved in the study.

Data Availability Statement: The datasets generated and/or analyzed during the current study are available from the corresponding author on reasonable request. All other public datasets used are available and cited in the references.

Acknowledgments: We would like to thank the editor and the reviewers for their valuable comments.

Conflicts of Interest: On behalf of all the authors, the corresponding author states that there are no conflicts of interest.

References

1. Kwon, Y.H.; Lobo, D.V. Age classification from facial images. In Proceedings of the IEEE Conference on Computer Vision and Pattern Recognition, Seattle, WA, USA, 21–23 June 1994; pp. 762–767.
2. Lanitis, A.; Taylor, C.J.; Cootes, T.F. Toward automatic simulation of aging effects on face images. *IEEE Trans. Pattern Anal. Mach. Intell.* **2002**, *24*, 442–455. [CrossRef]
3. Panis, G.; Lanitis, A.; Tsapatsoulis, N.; Cootes, T.F. Overview of research on facial ageing using the FG-NET ageing database. *IET Biom.* **2016**, *5*, 37–46. [CrossRef]
4. Yi, S.; Wang, X.; Tang, X. Deep Convolutional Network Cascade for Facial Point Detection. In Proceedings of the IEEE Internation Conference on Computer Vision and Pattern Recognition, Portland, OR, USA, 23–28 June 2013; pp. 3476–3483.
5. Taigman, Y.; Ming, Y.; Ranzato, M.; Wolf, L. Deep Face: Closing the Gap to Human-Level Performance in Face Verification. In Proceedings of the IEEE Conference on Computer Vision & Pattern Recognition, Columbus, OH, USA, 23–28 June 2014; pp. 1701–1708.
6. Ming, Y.; Zhu, S.; Lv, F.; Yu, K. Correspondence driven adaptation for human profile recognition. In Proceedings of the IEEE Conference on Computer Vision & Pattern Recognition, Colorado Springs, CO, USA, 20–25 June 2011; pp. 505–512.
7. Niu, Z.; Zhou, M.; Wang, L.; Gao, X.; Hua, G. Ordinal Regression with Multiple Output CNN for Age Estimation. In Proceedings of the IEEE Conference on Computer Vision & Pattern Recognition, Las Vegas, NV, USA, 27–30 June 2016; pp. 4920–4928.
8. Sendik, O.; Keller, Y. Deep Age: Deep Learning of face-based age estimation. *Signal Process. Image Commun.* **2019**, *78*, 368–375.
9. Chen, J.C.; Kumar, A.; Ranjan, R.; Patel, V.M.; Alavi, A.; Chellappa, R. A cascaded convolutional neural network for age estimation of unconstrained faces. In Proceedings of the International Conference on Biometrics Theory, Applications and Systems, Buffalo, NY, USA, 6–9 September 2016; pp. 20–28.
10. Zhang, K.; Gao, C.; Guo, L.; Sun, M.; Yuan, X.; Han, T.X.; Zhao, Z.; Li, B. Age Group and Gender Estimation in the Wild with Deep Ro R Architecture. *IEEE Access* **2017**, *5*, 22492–22503. [CrossRef]
11. Tao, W.; Zhao, Y.; Liu, L.; Li, H.; Xu, W.; Chen, C. A novel hierarchical regression approach for human facial age estimation based on deep forest. In Proceedings of the International Conference on Networking, Sensing and Control (ICNSC), Zhuhai, China, 27–29 March 2018; pp. 1512–1524.
12. Liu, H.; Lu, J.; Feng, J.; Zhou, J. Label-sensitive deep metric learning for facial age estimation. *IEEE Trans. Inf. Forensics Secur.* **2018**, *13*, 292–305. [CrossRef]
13. Taheri, S.; Toygar, O. Multi-stage age estimation using two level fusions of handcrafted and learned features on facial images. *IET Biom.* **2019**, *8*, 24–133. [CrossRef]
14. Li, K.; Zhang, G.; Wang, Y.; Wang, P.; Ni, C. Hand-dorsa vein recognition based on improved partition local binary patterns. In Proceedings of the 10th Chinese Conference (CCBR 2015), Tianjin, China, 13–15 November 2015; pp. 312–320.
15. Zhenghua, S.; Zhihua, X. Dorsal hand vein recognition based on transmission-type near infrared imaging and deep residual network with attention mechanism. *Opt. Rev.* **2022**, *29*, 335–342.
16. He, K.; Zhang, X.; Ren, S.; Sun, J. Deep residual learning for image recognition. In Proceedings of the IEEE Conference on Computer Vision and Pattern Recognition, Washington, DC, USA, 27–30 June 2016; pp. 770–778.
17. Zhu, Y.; Ni, X.; Yao, Y. Face Recognition Based on SVDF and Deep Learning with Attention. *J. Chin. Comput. Syst.* **2020**, *41*, 1763–1767.
18. Lee, J.H.; Chan, Y.M.; Chen, T.Y.; Chen, C.S. Joint estimation of age and gender from unconstrained face images using lightweight multi-task cnn for mobile applications. In Proceedings of the 2018 IEEE Conference on Multimedia Information Processing and Retrieval (MIPR), Miami, FL, USA, 10–12 April 2018; pp. 162–165.

19. Lin, L.E.; Lin, C.H. Data augmentation with occluded facial features for age and gender estimation. *IET Biom.* **2021**, *10*, 640–653. [CrossRef]
20. Wang, X.; Yili Hamu, Y. Face age estimation with dual-stream enhanced shallow network under unrestricted conditions. *Laser J.* **2022**, *43*, 88–94.
21. Li, J.; Li, K.; Zhang, G.; Wang, J.; Li, K.; Yang, Y. Recognition of Dorsal Hand Vein in Small-Scale Sample Database Based on Fusion of ResNet and HOG Feature. *Electronics* **2022**, *11*, 2698. [CrossRef]
22. Kuzu, R.S.; Maiorana, E.; Campisi, P. On the intra-subject similarity of hand vein patterns in biometric recognition. *Expert Syst. Appl.* **2022**, *192*, 116305. [CrossRef]
23. Cimen, M.E.; Boyraz, O.F.; Yildiz, M.Z.; Boz, A.F. A new dorsal hand vein authentication system based on fractal dimension box counting method. *Optik* **2021**, *226*, 165438. [CrossRef]
24. Huang, X.; Zhao, G.; Hong, X.; Zheng, W.; Pietikäinen, M. Spontaneous facial micro-expression analysis using spatiotemporal completed local quantized patterns. *Neurocomputing* **2016**, *175*, 564–578. [CrossRef]
25. Hanaoka, K.; Ngan, M.; Yang, J.; Quinn, G.W.; Hom, A.; Grother, P. *Face Analysis Technology Evaluation: Age Estimation and Verification*; National Institute of Standards and Technology: Gaithersburg, MD, USA, 2024. [CrossRef]
26. Yeom, H.-G.; Lee, B.-D.; Lee, W.; Lee, T.; Yun, J.P. Estimating chronological age through learning local and global features of panoramic radiographs in the Korean population. *Sci. Rep.* **2023**, *13*, 21857. [CrossRef] [PubMed]
27. Bekhouche, S.E.; Benlamoudi, A.; Dornaika, F.; Telli, H.; Bounab, Y. Facial Age Estimation Using Multi-Stage Deep Neural Networks. *Electronics* **2024**, *13*, 3259. [CrossRef]
28. Rasmus, R.; Radu, T.; Gool, L.V. DEX: Deep expectation of apparent age from a single image. In Proceedings of the IEEE International Conference on Computer Vision Workshop, Boston, MA, USA, 7–12 June 2015.

photonics

MDPI

Article

A Study on the Improvement of YOLOv5 and the Quality Detection Method for Cork Discs

Liguo Qu [1,2,*], Guohao Chen [3], Ke Liu [1] and Xin Zhang [1]

[1] School of Physics and Electronic Information, Anhui Normal University, Wuhu 241002, China; keliu@ahnu.edu.cn (K.L.); 2121011165@ahnu.edu.cn (X.Z.)
[2] Anhui Intelligent Robot Information Fusion and Control Engineering Research Center, Wuhu 241002, China
[3] Wuhan Mingke rail Transit Equipment Co., Ltd., Wuhan 430074, China; chen2711@ahnu.edu.cn
* Correspondence: lgqu@ahnu.edu.cn

Abstract: Combining machine vision and deep learning, optical detection technology can achieve intelligent inspection. To address the issues of low efficiency and poor consistency in the quality classification of cork discs used for making badminton heads, research on optimizing the YOLOv5 image-processing algorithm was conducted and applied to cork disc quality detection. Real-time images of cork discs were captured using industrial cameras, and a dataset was independently constructed. A GAN-based defect synthesis algorithm was employed to resolve the lack of defect samples. An attention mechanism was embedded in the YOLOv5 backbone network to enhance feature representation. The number of anchors in the YOLOv5 detection layer was reduced to address similar sample sizes, a center-matching strategy was designed to balance positive samples, and a shortest-distance label assignment algorithm was developed to eliminate ambiguities, improving accuracy and reducing postprocessing complexity. Detection results were integrated into quality classification. Experiments on the NVIDIA RTX3080 GPU demonstrated that the optimized algorithm improved the original YOLOv5 F1 score by 2.4% and mF1 score by 9.0%, achieving a quality classification F1 score of 95.1%, a processing speed of 178.5 FPS, and an mAP of 81.5%. Comparative experiments showed that the improved algorithm achieved the best detection accuracy on the cork disc dataset while maintaining high processing speed.

Keywords: deep learning; YOLOv5; defect detecting; cork disc; optical detection technology

1. Introduction

Natural cork, a product derived from the bark of Mediterranean oak trees, is extensively utilized due to its excellent elasticity, sealing properties, and abrasion resistance. It is widely applied in the manufacturing of premium badminton shuttlecocks and cork stoppers for sealing wine bottles, possessing significant commercial value [1]. In the production of badminton shuttlecock heads, entire sheets of cork bark are cut into small cork discs. The quality screening of these cork discs is currently one of the crucial processes in the manufacturing of badminton shuttlecock heads. However, natural cork, influenced by environmental factors during its growth, exhibits complex surface defects and varied textures, making it impossible to find two samples with identical defect patterns. Traditional manual visual inspection methods are highly subjective, influenced by factors such as physical fatigue and work experience, resulting in inconsistent sorting quality. Therefore, achieving stable quality control presents substantial challenges.

Research on the quality screening of cork discs based on optical detection technology was initiated in the 1990s. Cork images were captured using cameras to conduct automatic visual inspection of cork-related products. In 1997, Chang et al. [2] designed a feature extraction method involving morphological filtering, contour extraction, and tracking. A complex neural network was used as a classifier in the cork stopper quality classification system, achieving the classification of eight different quality grades of cork stoppers. In

2000, Gonzalez-Adrados et al. [3] proposed a cork board quality grading system, which was based on the analysis of dozens of data features from cross-sectional and tangential images of cork boards, identifying three different types of defects. Discriminant analysis was further employed for quality grading, with classification results surpassing manual classification. Costa et al. [4–6] analyzed the contribution of each porosity feature to cork stopper grading and developed a cork stopper quality classification system based on canonical discriminant analysis and stepwise discriminant analysis techniques, achieving a 14% error rate for the surface classification of seven standard commercial quality grades of cork stoppers. In 2009, Georgieva et al. [7] studied an intelligent machine vision system for the classification of seven different types of cork bricks. Feature generation was performed using Laws' masks, and the feature vectors were processed using linear discriminant analysis and principal component analysis for cork brick classification. In 2010, Paniagua et al. [8] constructed a cork stopper classification vision system that used a static threshold method to determine defect areas and morphological calculations to measure defect sizes, classifying cork stoppers using a neuro-fuzzy classifier. In 2015, Vanda Oliveira et al. [9] characterized the surface porosity of cork stoppers of three grades using image analysis methods and established a predictive classification model using stepwise discriminant analysis, achieving a classification accuracy of 75%. Other advanced techniques have also been applied to cork classification, such as neutron radiography and tomography for analyzing internal defects of cork, and volatile organic compound (VOC) analysis for natural cork stoppers with different porosity levels [10]. These methods generally involve two main steps: first, feature extraction from cork images using texture feature generation and extraction techniques, and then classification using neural networks.

In recent years, automatic optical detection technology based on deep learning has been widely and importantly applied in intelligent manufacturing for tasks such as positioning detection and surface quality inspection of components [11,12]. Examples include rubber threading line detection [13], metal surface defect detection [14–16], fabric defect detection [17], and sanitary ceramic defect detection [18], providing new solutions for cork disc quality screening. Machine vision combined with automated equipment enables automatic detection systems, which offer advantages such as accuracy, efficiency, and continuous operation. Since the introduction of AlexNet [19] in 2012, numerous excellent deep-learning algorithms have emerged, including R-CNN [20–22], ResNet [23], SSD [24], RetinaNet [25], CenterNet [26], and ConvNeXt [27]. In industrial applications, the real-time object detection algorithm YOLO (You Only Look Once) has become one of the most popular deep-learning detection frameworks. YOLO [28–30], proposed by Redmon, has since been integrated with the latest object detection algorithms by various researchers, leading to the development of efficient detection algorithms such as YOLOv4 [31], YOLOv5 [32], YOLOX [33], YOLOv6 [34], and YOLOv7 [35]. Among these, YOLOv5 has garnered extensive attention due to its optimal balance between accuracy and speed. YOLOv5 maintains high accuracy while offering rapid detection speeds, and it allows for model customization according to user-specific requirements. With a compact model size, it is particularly suited for edge computing and deployment environments with limited resources. Therefore, the application of YOLOv5 to the quality inspection of cork discs can better facilitate the automatic extraction of cork defect features and the implementation of a low-cost system. This study addresses the issues of low efficiency and poor consistency in cork disc quality screening by investigating an improved YOLOv5-based method for cork disc quality detection, achieving rapid and precise cork disc quality inspection.

2. Materials

2.1. Dataset Construction

The cork discs in the dataset were sourced from a badminton production factory. Images of each cork disc's front and back sides were captured using industrial cameras. As shown in Figure 1a, an example image of a cork disc is presented. After preliminary processing, the oak bark is segmented into cork discs with a diameter of approximately

27 mm and a height of about 5 mm. Each shuttlecock head is formed by adhering three cork discs together. The top cork disc, which requires the highest quality, has holes punched around it to insert feathers. Cork discs of slightly lower quality are placed in the middle or bottom layers.

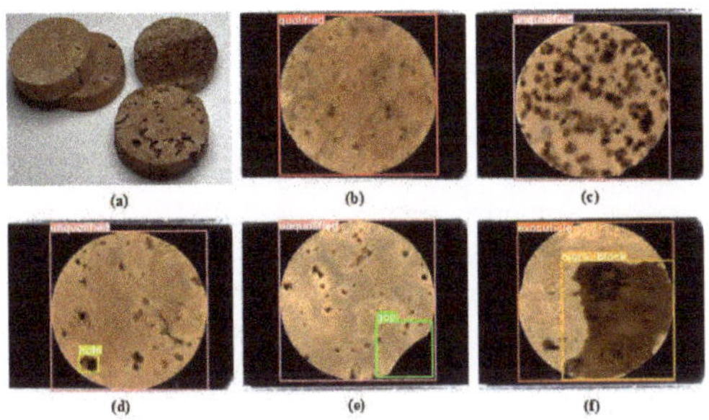

Figure 1. Cork discs and classification.(**a**) Example images of cork disc. (**b**) Qualified cork disc. (**c**) Unqualified cork disc with many holes. (**d**) Unqualified cork disc with a large hole in the edge. (**e**) Unqualified cork disc with a notches in the edge. (**f**) Unqualified cork disc with black spots.

In this study, both quality classification and defect detection tasks for cork discs are conducted simultaneously. Quality classification is divided into three categories: qualified, unqualified, and outer bark; defect detection is divided into three categories: holes, notches, and black spots. In Figure 1c, holes are annotated. Pores in the oak bark appear as black spots of varying sizes on the disc's surface, but only holes larger than 2 mm are annotated as hole defects, as small holes do not affect the shuttlecock head's quality. In Figure 1c, notches are annotated, indicating missing portions at the edge of the cork disc, primarily caused by natural factors or during cutting. In Figure 1f, black spots are annotated, originating from the oak bark's outer cuticle, which is hard, non-elastic, and of no use. Regarding these defects, cork discs can be annotated with different quality classifications. In Figure 1b, a qualified disc is annotated, indicating a surface with minimal defects, suitable for the upper layer of the shuttlecock head to insert feathers. In Figure 1c,e, unqualified discs are annotated, indicating they cannot be used for the upper layer but can be used for the middle and lower layers. In Figure 1f, the outer bark is annotated, which is of no use. Figure 1d appears to have few surface defects but is annotated as unqualified due to a hole defect located in the 1–4 mm edge region (feather insertion area). Although large holes in the center area may still be classified as qualified and used for the upper layer, the classification of hole defects requires further determination of hole size and distribution for precise quality classification.

To train the model, a total of 8570 cork images were collected. After preprocessing, each sample had a pixel size of 480 × 480. The images were annotated using the LabelImg tool, and the annotation results were saved in XML format. All images were randomly divided at a ratio of approximately 8:1:1, constructing an original cork disc dataset that includes 6770 training images, 900 validation images, and 900 test images.

2.2. Data Augmentation

In the cork disc dataset, the proportion of defective samples is small, resulting in an imbalance in the training data and poor generalization ability of the model. Collecting a large number of such samples is challenging; therefore, a defect synthesis algorithm based on Generative Adversarial Networks (GANs) is proposed to address the issue of insufficient defective samples. Due to the varying shapes of naturally occurring hole defects, a GAN is employed to generate new hole defects to simulate this variability. The GAN consists of two modules: a generator and a discriminator. Its main objective is to obtain the optimal solution to the following objective function [36], as shown in Equation (1).

$$\min_{G}\max_{D} V(D,G) = E_{x\sim p_{data}(x)}[\log D(x)] + E_{z\sim p_z(z)}[\log(1 - D(G(z)))] \quad (1)$$

In Equation (1), x represents the training images, G represents the generator, D represents the discriminator, and z represents the generated images. During the adversarial process, the generator creates fake data samples and attempts to deceive the discriminator, while the discriminator tries to distinguish between real and fake samples. When the discriminator cannot determine the source of the images, it can be considered that the generator is capable of producing images that follow the same distribution as the training set.

Fake "holes" similar to those in the original dataset are generated using a Generative Adversarial Network (GAN). These fake holes are integrated as new defects into cork discs to create new samples. Figure 2 illustrates the proposed data augmentation (DA) algorithm process, which includes the following steps:

(1) Selection of Background Images: "Qualified" samples from the original dataset are selected as background images. To prevent data repetition, each background image undergoes horizontal or vertical flipping.
(2) Transformation of Defect Images: Defect images generated by the GAN are randomly selected and subjected to affine transformations, including random horizontal flipping, vertical flipping, scaling, and rotation.
(3) Binarization and ROI Extraction: The transformed defect images are binarized using the OTUS method to obtain the Region of Interest (ROI) mask, identifying the defect ROI.
(4) Background Region Extraction: A center point is randomly selected within a specific area of the cork disc background image (1–4 mm from the outer diameter where holes are punched for feather insertion). A background region of the same size as the transformed defect image is cropped around this center point.
(5) Defect Integration: The defect ROI is fused with the background region and overlaid onto the cork disc background image.

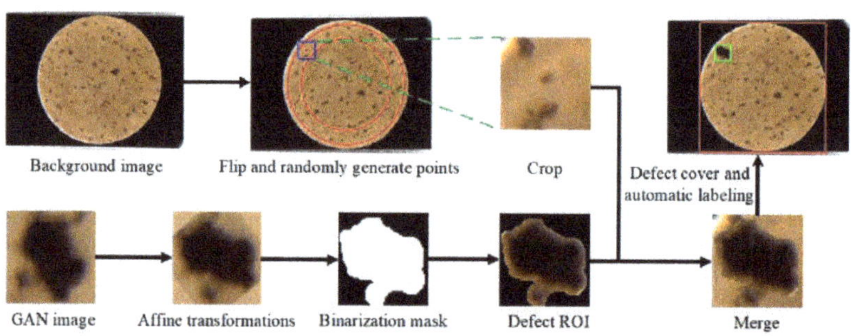

Figure 2. Sample defect synthesis algorithm based on GAN.

The GAN-based defect synthesis algorithm can generate a large number of hole defects. However, augmenting notch defects is challenging due to their occurrence only at the edges of cork discs and their directional nature; thus, random angle rotation is applied only to samples with notches. For black spot defects, due to their distinctive features and high detection accuracy, data augmentation is not performed. For hole and notch defects, 500 new samples are generated for each type. These samples are randomly divided into training, validation, and test sets at approximately an 8:1:1 ratio. After data augmentation, the dataset contains 7570 training images, 1000 validation images, and 1000 test images.

3. Quality Detection Method of Cork Discs Based on Improved YOLOv5

3.1. Overall Architecture

The overall architecture of the cork disc quality detection model is illustrated in Figure 3. The model architecture consists of three parts: the backbone network, the neck, and the detection head. It is an improvement upon YOLOv5, with the primary modifications including the integration of a Convolutional Block Attention Module (CBAM) into the backbone network to further enhance the model's feature representation capability. Given the uniform size of the cork discs, the number of prediction anchors in the detection layer has been reduced. A Center Match (CM) strategy is introduced to expand the range of positive sample selection, thereby balancing the number of positive samples. A shortest-distance label assignment (SDLA) strategy is proposed to address the issue of ambiguous sample regression. Additionally, a Detection Result Processing (DRP) algorithm is designed to further improve accuracy by aligning with the quality screening requirements for badminton cork discs.

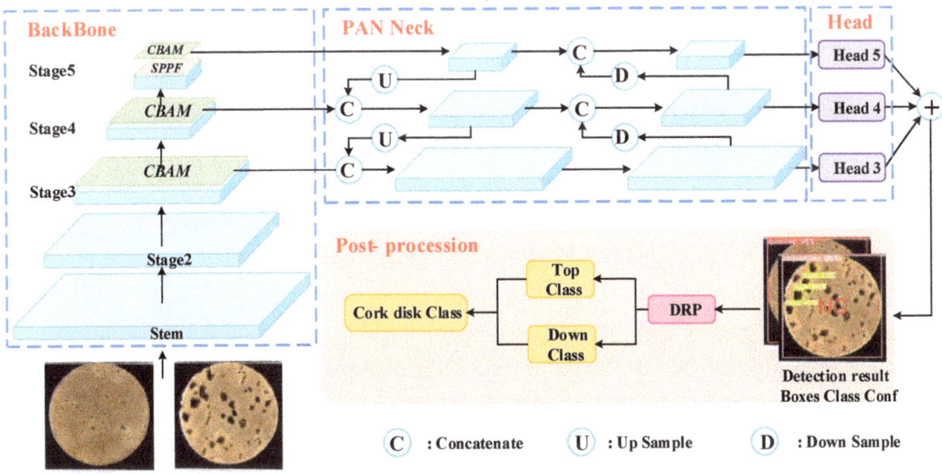

Figure 3. Architecture of the quality inspection model for cork discs.

In the backbone network section, the 6 × 6 convolution in the Stem module provides a larger receptive field spatially, enabling the acquisition of richer image features. Down-sampling convolutions expand the channel dimensions while reducing the feature map scale, segmenting image features into different stages. The feature extraction structures at each stage combine the advantages of cross-stage local network structures and bottleneck structures [23], reducing computational cost while enhancing the backbone's feature extraction capability. The CBAM is embedded in the backend of Stage 3, Stage 4, and Stage 5 of the backbone network, as a large amount of high-level semantic information is present in the deeper layers of the backbone network. CBAM improves the network's capability to express features.

The neck of YOLOv5 consists of an improved Path Aggregation Network (PAN) structure, which aggregates multi-scale features by connecting low-level physical features with high-level semantic features through bottom-up paths, thereby constructing pyramid feature maps and providing feature inputs for the detection head. The detection head performs predictions on the three scale feature maps generated by the neck, using a lightweight 1 × 1 convolutional layer. For quality detection, both sides of the cork disc are simultaneously captured and preprocessed to a size of 480 × 480. The two images are stacked into a batch of 2 and input into the neural network. Subsequently, the improved YOLOv5 performs data inference and detects both images simultaneously.

3.2. CBAM

To further enhance the feature expression capability of the model, the CBAM is embedded in the backbone network based on YOLOv5. The CBAM is a lightweight convolutional attention module that operates on both channel and spatial information. It consists of two submodules: the channel attention module (CAM) and the spatial attention module (SAM), as illustrated in Figure 4. The two modules are connected in series and introduce parallel branches relative to the input feature path. This configuration allows the generation of attention feature map information sequentially along both the channel and spatial dimensions, thereby enabling adaptive feature refinement.

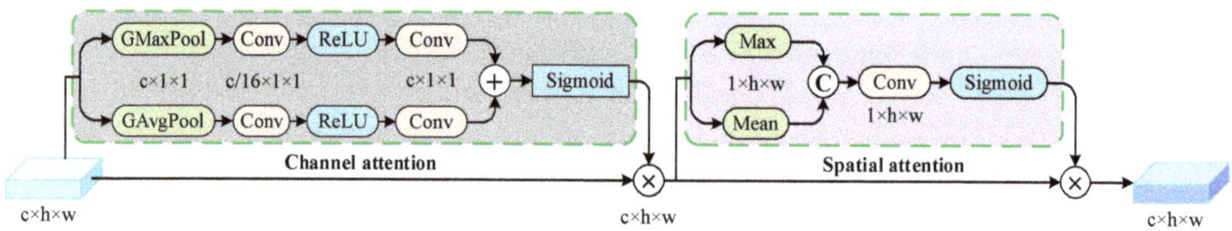

Figure 4. CBAM structure diagram.

The computation of the CBAM is described by Equation (2).

$$F' = M_c(F) \otimes F$$
$$F'' = M_s\left(F'\right) \otimes F' \quad (2)$$

In the formula, $F \in \mathbb{R}^{C\times H\times W}$ represents the input features, while $M_c \in \mathbb{R}^{C\times 1\times 1}$ and $M_s \in \mathbb{R}^{1\times H\times W}$, respectively, denote the channel attention features and spatial attention features. The symbol $\otimes$ signifies element-wise multiplication. The computation within the channel attention module is illustrated as shown in Equation (3).

$$M_c = \sigma(Conv_2(ReLU(Conv_1(F^c_{max}))) + Conv_2(ReLU(Conv_1(F^c_{avg})))) \quad (3)$$

In the formula, $F^c_{max} \in \mathbb{R}^{C\times 1\times 1}$ and $F^c_{avg} \in \mathbb{R}^{C\times 1\times 1}$ represent the channel weights obtained through the application of global maximum pooling and global average pooling operations across all channels, respectively. σ denotes the sigmoid activation function, and $Conv_1 \in \mathbb{R}^{C\times \frac{C}{r}\times 1\times 1}$, $Conv_2 \in \mathbb{R}^{\frac{C}{r}\times C\times 1\times 1}$, r is a dimensionality reduction factor employed to decrease computational load, with a default setting of 16, while the minimum value of $\frac{C}{r}$ is set to 8. The computation of the spatial attention mechanism module is expressed as shown in Equation (4).

$$M_s = \sigma(Conv^{7\times 7}(Cat(F^s_{max}, F^s_{avg}))) \quad (4)$$

In the equation, $F^s_{max} \in \mathbb{R}^{1\times H\times W}$ and $F^s_{avg} \in \mathbb{R}^{1\times H\times W}$ represent two-dimensional spatial mappings obtained from two types of pooling operations. *Cat* signifies the concatenation operation, and $Conv^{7\times 7} \in \mathbb{R}^{2\times 1\times H\times W}$ represents a convolution operation with a 7×7 kernel.

The CBAM is embedded at the backend of Stage 3, Stage 4, and Stage 5 because deep layers of the backbone network are replete with abundant high-level semantic information. Utilizing attention modules for information fusion in these areas allows the model to differentiate features more effectively.

3.3. CM Strategy

YOLOv5 is an anchor-based detector that sets three different scales of anchors for predictions at each detection layer. While multi-scale prediction can offer performance improvements, it results in slower prediction speed and higher postprocessing complexity. Additionally, given that the target scales of cork disc objects in the dataset are nearly uniform, the benefits of multi-scale prediction are minimal. Therefore, the number of anchors is set to one to reduce the number of predictions and enhance processing speed. The reduction in the number of anchors leads to a decrease in the number of positive samples used for training. To address this, a CM strategy is employed to expand the range of positive sample selection and balance the number of positive samples.

As shown in Figure 5a, the positive sample selection strategy of YOLOv5 matches predictions from three grids with the ground truth targets. During the selection process, when the center of a ground truth target falls within a particular grid, that grid is chosen. When the number of anchors is reduced to one, the number of selected positive samples decreases from nine to three, resulting in a significant reduction in the number of positive samples, which in turn lowers the algorithm's convergence rate and accuracy. To balance the number of positive samples, a Center Match (CM) strategy is proposed: predictions from the grid containing the center of the ground truth target and its eight neighboring grids are selected as positive samples, as shown in Figure 5b. After replacing the original positive sample selection strategy with the Center Match strategy, the transformation formula for the prediction results relative to the ground truth bounding box center coordinates changes. The calculations for YOLOv5 with the Center Match strategy are given by Formula (5).

$$\begin{aligned} B_{x,y} &= \left(2 \times \sigma\left(P_{x,y}\right) - 0.5\right) + G_{x,y} \quad for\, YOLOv5 \\ B_{x,y} &= \left(3 \times \sigma\left(P_{x,y}\right) - 1.0\right) + G_{x,y} \quad for\, CM \end{aligned} \tag{5}$$

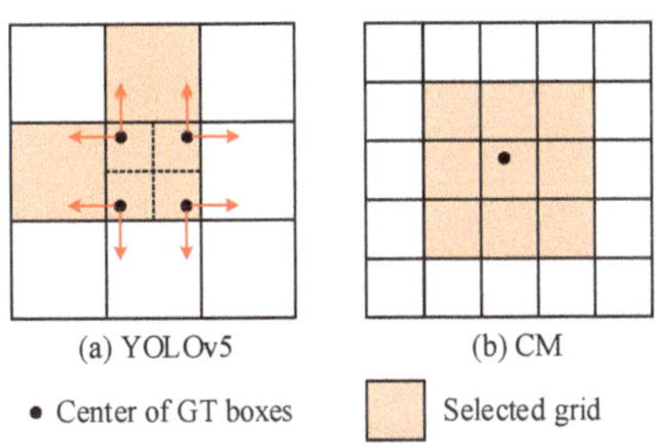

Figure 5. Schematic diagram of YOLOv5 and CM positive sample selection strategy. The grid is divided into four regions by using dotted cross lines. If the center point is predicted within a certain region, the two adjacent grids are selected according to the direction of the arrows.

In the formula, σ represents the sigmoid function, $\sigma\left(P_{x,y}\right)$ denotes the predicted offsets for x and y, and $G_{x,y}$ signifies the coordinates of the top-left corner of the grid point in the x–y plane.

Compared to the original method of selecting positive samples, the CM strategy maintains the same quantity of positive samples while reducing the number of model prediction outputs, enhancing the model's prediction speed and reducing postprocessing complexity. Moreover, the positive samples involved in loss calculation come from the target's central area, where prediction results are generally of higher quality. Calculating loss with higher-quality prediction results enriches the gradient information.

3.4. SDLA Strategy

When selecting positive samples, sufficient proximity between targets can lead to a challenging ambiguity: predictions from certain grids may be selected as positive samples for multiple ground truth targets, resulting in confusion about which target the prediction should be regressed to. Such overlapping samples are referred to as ambiguous samples, as illustrated by the green striped grids in Figure 6.

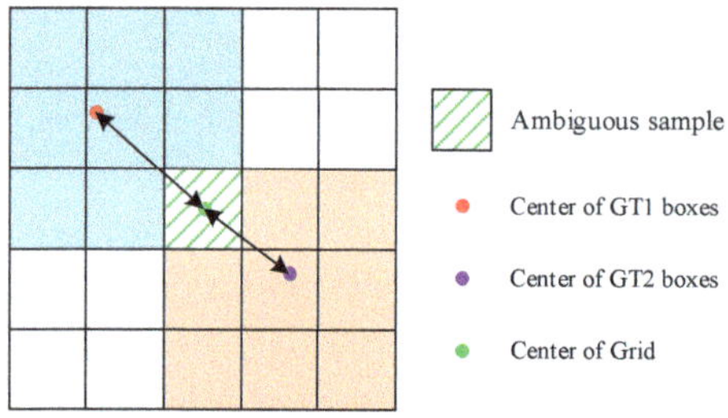

Figure 6. Schematic diagram of the SDLA strategy. The line with the arrow is the Euclidean distance from the anchor center to the truth targets.

YOLOv5 does not address such cases. It is noted that selecting positive samples within a smaller spatial range can significantly reduce the occurrence of ambiguous samples. The CM strategy, however, expands the positive sample selection space, increasing the likelihood of ambiguous samples, which impacts accuracy to some extent. Therefore, a simple and effective SDLA strategy is proposed to eliminate this ambiguity. The process is detailed in Figure 6. When an ambiguous sample is assigned to two ground truth targets, GT1 and GT2, the center point of the grid cell is first calculated, equal to the top-left corner coordinates plus 0.5 times the downsampling stride. The Euclidean distance from this center point to the centers of the two ground truth targets is then computed, and the target with the shortest distance is chosen as the regression target. As shown in Figure 6, the ambiguous sample is ultimately assigned to GT2. The SDLA strategy can effectively and simply eliminate erroneous gradient information during backpropagation.

3.5. DRP Algorithm

In the cork disc quality detection results, predictions for multiple categories are included. However, only one classification prediction for the cork disc is desired. Therefore, during non-maximum suppression of redundant detection boxes, the predictions for the three quality categories of the cork discs are processed together, retaining only the category with the highest confidence score. This approach suppresses multiple prediction categories and reduces the complexity of subsequent processing. However, for corks similar to those shown in Figure 1d, where the probability of model inference errors is higher, the Detection Result Processing (DRP) algorithm is employed to correct the final output category labels based on the relationship between cork disc categories and surface defects, thereby further enhancing classification accuracy.

The logic of the Detection Result Processing (DRP) algorithm is illustrated in Figure 7. Based on quality classification, the results are further refined by incorporating defect detection outcomes. The "black spot" defect is considered the most severe; cork discs with this defect are classified as the "outer skin" category. Therefore, when the category is "outer skin", the result is directly output without any further processing. For other categories, it is necessary to check for the presence of the "black spot" defect. If it is present, the output category will be changed to "outer skin".

A more complex scenario arises when the cork disc is classified as "qualified" but contains defects. In this case, the following steps are sequentially assessed: First, determine if the "black spot" defect is present. If it is, the category will be modified to "outer skin". If not, check for the presence of "notch" defects. If present, the category is changed to "unqualified".

Finally, for hole defects, classification will be based on three criteria: (1) whether the holes are located within a specific area, (2) whether the largest hole is greater than 4 mm in size, and (3) whether the total number of holes exceeds 3. If any of these conditions are met, the category will be updated to "unqualified".

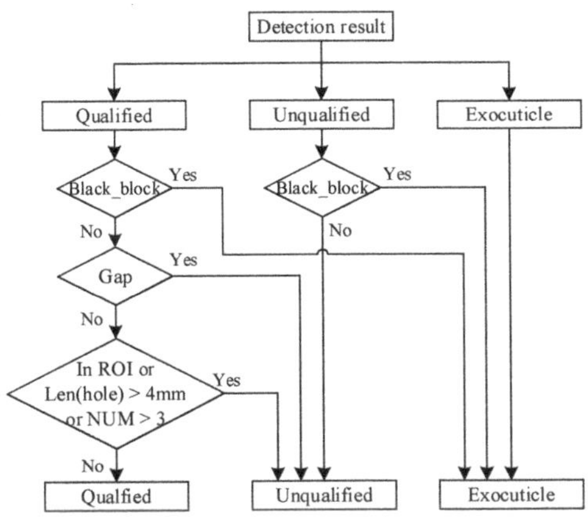

Figure 7. DRP logic diagram.

4. Experiment and Result Analysis

4.1. Experimental Environment and Evaluation Index

The hardware configuration for model training and testing includes an Intel Core™ i7-11700K processor and an NVIDIA RTX3080 GPU with 10 GB of memory. The above hardware platform comes from Lenovo Group Co., Ltd. in Beijing, China. The computer's operating system is Ubuntu 18.04, and the software environment is configured with Python 3.8, Pytorch 1.8, Opencv 4.1.2, and CUDA v11.1.

During model training, the number of training epochs is set to 300, with a batch size of 32. K-means clustering is used to recalculate anchors suitable for the cork disc dataset. For the original YOLOv5 model, anchors are set as follows: [[49, 49], [61, 76], [114, 46], [66, 148], [154, 98], [180, 205], [332, 226], [229, 328], [443, 441]]. After applying the CM strategy, anchors are set as follows: [[66, 66], [240, 237], [442, 441]]. Other hyperparameters are evolved using a genetic algorithm (GA) and selected as follows: the initial learning rate is 0.01, the momentum is 0.946, the weight decay is 0.00047, the bounding box regression loss gain is 0.053, the classification loss gain is 0.86, the classification binary cross-entropy loss positive weight is 0.82, the objectness loss gain is 0.566, and the objectness binary cross-entropy loss weight is 0.949.

For model evaluation, the confidence threshold is set to 0.45 and the IoU threshold to 0.7. The evaluation metric chosen for the detection model is the *F*1 score, as shown in Equation (6), because it considers both the *precision* and *recall* of the classification model, providing a more accurate reflection of the model's classification capability.

$$F1 = 2 \times \frac{precision \times recall}{precision + recall} \quad (6)$$

It is noteworthy that the model task includes both quality classification and defect detection. Therefore, in addition to using the mean F1 score (mF1) metric, we also consider the classification detection F1 score (CDF1) and the defect detection F1 score (Defect_F1). The CDF1 score determines the accuracy of the quality classification, while the defect detection results are used by the DRP algorithm to correct the quality classification.

4.2. Ablation Experiment

To verify the effectiveness of the algorithm improvements, ablation experiments were conducted on different modifications made to the YOLOv5 base network.

4.2.1. Effects of Model Size and Pre-Training Weights

YOLOv5 provides models of various scales along with corresponding pre-trained weights to meet the needs of different application scenarios. Generally, larger models offer better performance but slower inference speeds. To determine the most suitable model scale, experiments evaluated three models (n, s, and m) on the cork disc dataset with and without pre-trained weights, assessing mF1 scores and inference speeds. The results are shown in Figure 8. In Figure 8, the blue solid line and red dashed line represent the detection results of models with and without pre-trained weights, respectively. It is evident that using pre-trained weights improves detection performance across various model scales, as it better initializes model parameters. Furthermore, while larger models provide limited improvement in mF1 scores, inference time increases rapidly. Among them, the YOLOv5s model, with an inference speed of 4.5 ms and a detection mF1 score of 85.7%, achieves the best balance between speed and accuracy. Therefore, the YOLOv5s model is selected as the baseline model, and pre-trained weights are used to initialize parameters in subsequent training.

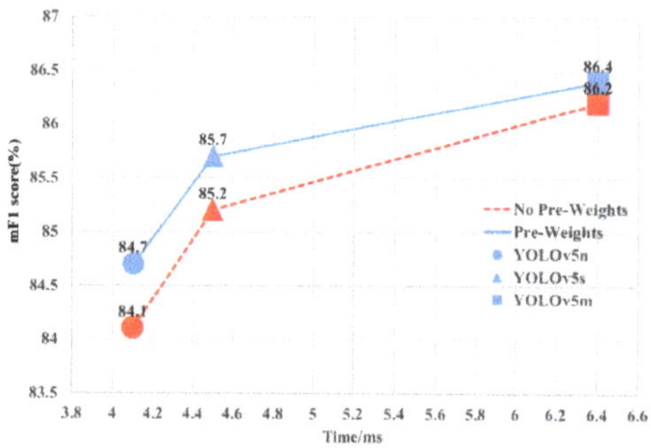

Figure 8. Detection speed and mF1 score of different YOLOv5 models. The circle represents YOLOv5n, the triangle represents YOLOv5s, and the square represents YOLOv5m, regardless of color.

4.2.2. Impact of Adding Defect Training

The core task of model inference is the quality classification of cork discs. The model was designed as both a dedicated quality classification model and a unified model (performing both quality classification and defect detection). In the dataset with data augmentation, the dedicated quality classification model achieved a CDF1 score of 91.1%, whereas the unified model achieved a CDF1 score of 93.8%, representing an improvement of 2.7%. This significant improvement is attributed to the inherent relationship between cork disc categories and defects. The neural network, through extensive data learning, gradually recognizes the relationship between them and considers the spatial activation features of defects when making classification predictions. Further explanation using the visual feature maps shown in Figure 9 reveals that these feature maps are derived from the Head 5 layer of the detection model. In Figure 9, the detection model without defect targets can only learn global features, while the feature maps with defect detection exhibit strong spatial feature activation at defect targets (e.g., holes in the lower-left corner), leading to more accurate classification judgments.

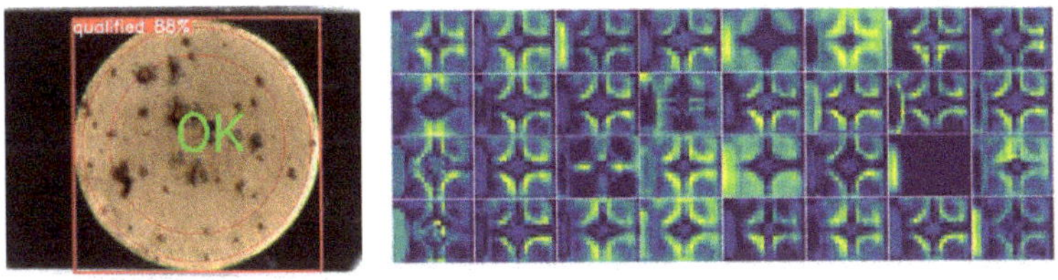

(a)

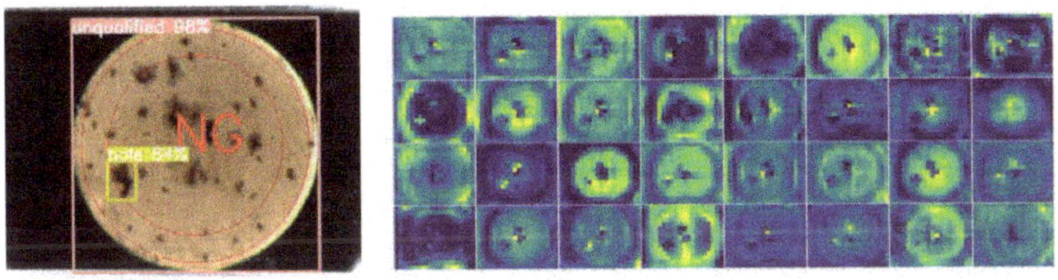

(b)

Figure 9. Removal and addition of visual feature maps of defect target models. (**a**) The model output feature maps with removed defect sample training. (**b**) The model output feature maps with added defect sample training.The area between the two red circles represents the 1-4mm region of interest.

4.2.3. Impact of DA and DRP

The amount of training data directly affects the final performance of the detection model. Data augmentation methods were employed to add 500 "hole" and "gap" defects to the dataset to mitigate the issue of imbalanced defect samples.

As shown in Table 1, experimental results indicate that when the number of defect samples is low, the Defect_F1 score for the three types of defects is only 62.7%. After using DA to increase the number of defect samples, the Defect_F1 score significantly improved by 14.1%, making the model's defect detection more robust. Additionally, the CDF1 score also increased slightly by 1.1%. This improvement is partly due to the model's enhanced ability to extract defect features, which ultimately impacts the overall classification judgment. Although the unified model improves generalization and efficiency through shared lower-level features and joint training, this self-learned cognition is not perfect, and errors may still occur in some cases. Therefore, using the DRP algorithm to correct the quality classification based on defect detection results further enhances classification accuracy, with the CDF1 score increasing by 0.7% from 93.8%.

Table 1. The effect of DA data enhancement was compared.

Model	CDF1	Defect_F1	mF1
YOLOv5	92.7%	62.7%	77.7%
YOLOv5 + DA	93.8%	76.8%	85.3%
YOLOv5 + DA + DRP	94.5%	76.8%	85.7%

4.2.4. Impact of CM and SDLA

The performance of both the CM and SDLA optimization methods was evaluated in the experiment. These methods are "bag-of-freebies" techniques focused on optimizing the training process [31], aimed at improving object detection accuracy by increasing training costs without adding to the inference cost.

Table 2 compares the impact of the two methods on model accuracy and training time. Table 3 analyzes the proportion of ambiguous samples among all positive samples when different positive sample selection strategies are applied, with data from YOLOv5 tested using an anchor count of one.

Table 2. The impact of CM and SDLA strategies on accuracy and training time.

Model	CDF1	mF1	Train (Time/h)
YOLOv5 + DA + DRP	94.5%	85.7%	5.61
YOLOv5 + DA + DRP + CM	94.6%	86.4%	5.61
YOLOv5 + DA + DRP + CM + SDLA	94.9%	86.4%	6.50

Table 3. The proportion of fuzzy samples to all positive samples under different methods.

Method	Amb Samp	Pos Samp
YOLOv5	599	142,875
CM	3824	428,625
CM + SDLA	0	428,625

"Amb samp" denotes ambiguous samples, and "Pos samp" denotes positive samples.

Table 2 shows that the CM strategy slightly increased the CDF1 score by 0.1% and the mF1 score by 0.7% without any additional training cost. However, as indicated in Table 3, the CM strategy expanded the positive sample selection range, causing the number of ambiguous samples in the training process to increase from 599 to 3824, effectively doubling the probability of ambiguous samples, which destabilized the training process.

The SDLA strategy, which applies a simple shortest spatial distance principle to handle ambiguous samples, reduced the number of ambiguous samples to zero, thereby eliminating erroneous gradient information during training. This led to a 0.3% increase in the CDF1 score. However, the SDLA strategy requires processing on individual images, extending the training time by 0.89 h. Additionally, repeated training of models using the SDLA strategy yields fully consistent training results. This indicates that the SDLA strategy thoroughly eliminates ambiguous samples, ensuring complete consistency in data gradient information under the same training images and parameters.

4.2.5. Impact of CBAM

The experimental results of inserting attention modules at different positions in the YOLOv5 + DA + DRP + CM + SDLA model backbone network are shown in Table 4. CBAMn represents the placement of n CBAMs in the backbone network following a top-down order (Stage 5, Stage 4, Stage 3, Stage 2). Latency refers to the inference delay caused by CBAMn, excluding preprocessing and postprocessing times.

Table 4. Impact of CBAM on performance.

CBAM	Latency	CDF1	mF1
CBAM $\times$ 1	5.0 ms	94.1%	86.3%
CBAM $\times$ 2	5.3 ms	93.3%	84.9%
CBAM $\times$ 3	5.6 ms	95.1%	86.7%
CBAM $\times$ 4	5.9 ms	94.4%	86.6%

The experiment demonstrates that as the number of inserted attention modules increases, the model's inference delay also increases, with each additional CBAM contributing an extra 0.3 ms to the inference delay. According to the results in Table 4, embedding three CBAM attention modules at the end of Stage 5, Stage 4, and Stage 3 achieves the best balance, with an inference time of 5.6 ms and an mF1 score of 86.7%.

4.2.6. Accuracy Analysis

Table 5 summarizes the impact of different optimization methods on detection accuracy. Sequential application of DA, the DRP algorithm, the CM strategy, the SDLA strategy, and the CBAM attention mechanism to the YOLOv5s baseline model led to incremental improvements in model performance. On the cork disc dataset, a total improvement of 2.4% in CDF1 score and 9.0% in mF1 score was achieved, resulting in a final optimized detection algorithm with a CDF1 score of 95.1% and an mF1 score of 86.7%.

Table 5. The impact of different optimization methods on model precision.

Model	CDF1	mF1
(1) YOLOv5	92.7%	77.7%
(2) YOLOv5 + DA	93.8%	85.3%
(3) YOLOv5 + DA + DRP	94.5%	85.7%
(4) YOLOv5 + DA + DRP + CM	94.6%	86.4%
(5) YOLOv5 + DA + DRP + CM + SDLA	94.9%	86.4%
(6) YOLOv5 + DA + DRP + CM + SDLA + CBAM	95.1%	86.7%

Additionally, further analysis revealed that the DRP algorithm effectively improved classification accuracy without affecting inference time, though it did increase postprocessing time by an average of 1.5 ms per detection. The CM strategy reduced inference time by 2.3 ms and postprocessing time by 1.1 ms, as fewer predictions resulted in faster prediction speeds and lower postprocessing complexity. The SDLA strategy enhanced the model's classification accuracy without affecting inference speed, as it only increased training costs. The CBAM improved detection accuracy but increased inference time due to the additional parameters and computational load introduced to the model backbone network. Despite the increased inference time, CBAM's contribution to accuracy improvement was more significant.

The results of the optimized model (Model 6 in Table 5) compared to the original YOLOv5 model are shown in Figure 10. For five typical samples selected from the actual test results, the improved model demonstrated enhanced detection accuracy, particularly for samples 3, 4, and 5, where the original YOLOv5 model made incorrect quality classifications. The improved YOLOv5 model correctly performed quality classification and accurately identified defect features.

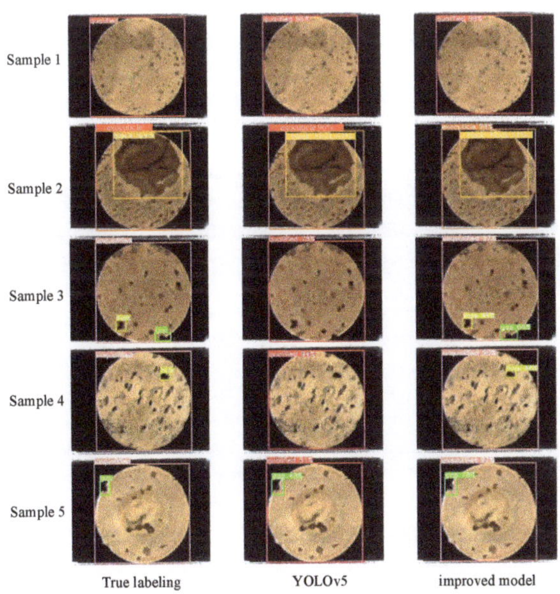

Figure 10. Comparison of detection results between YOLOv5 and the improved model.

4.3. Comparison Experiments

To validate the advancement of the improved algorithm, a comparative experiment was conducted with six popular object detection network models: Faster RCNN, RetinaNet, CenterNet, YOLOX, YOLOv4, and YOLOv7. Identical training parameters were set for all models, and their performance was tested on the same cork disc test dataset, as shown in Table 6. The single-stage detector YOLO models were able to process images at speeds exceeding 100 FPS and achieved an mF1 score of no less than 80% on the cork disc test dataset. In contrast, the classic models Faster RCNN and RetinaNet exhibited lower performance in both accuracy and speed, while the keypoint-based CenterNet demonstrated faster inference speed but lower detection accuracy.

Table 6. Comparison of different object detection models.

Model	Platform	Backbone	Type	mF1	mAP0.5	mAP	FPS
Faster RCNN	MMDetection	ResNet50	Anchor-based	77.2%	79.9%	71.9%	57.7
RetinaNet	MMDetection	ResNet18	Anchor-based	78.6%	78.6%	70.9%	88.6
CenterNet	MMDetection	ResNet18	Keypoint-based	74.9%	78.2%	69.9%	194.3
YOLOX	MMDetection	Darknet	Anchor-free	85.2%	79.7%	72.6%	102.8
YOLOv4	Darknet	Darknet	Anchor-based	75.0%	82.2%	58.7%	326.7
YOLOv7	YOLOv7	E-ELAN	Anchor-based	82.9%	77.4%	72.0%	227.2
Ours	YOLOv5	Darknet	Anchor-based	86.7%	89.4%	81.5%	178.5

In the YOLO series models, YOLOv7 claims to surpass YOLOv5 in performance. However, the results in Table 6 demonstrate that our improved YOLOv5 achieves the best balance between accuracy and speed, with an mF1 score of 86.7% and an mAP of 81.5%, surpassing all other detection models listed in terms of cork disc classification accuracy. Additionally, it processes at a speed of 178.5 FPS, which, while slightly lower than YOLOv7, still meets practical requirements.

5. Conclusions

Through in-depth research and experimental validation, a method for cork disc quality detection based on the improved YOLOv5 has been successfully developed. A proprietary cork disc dataset was constructed, including 8570 original images and 1000 augmented images, created using data augmentation algorithms. Tailored to the characteristics of cork disc detection and combined with the YOLOv5 model, a series of deep-learning optimization methods were proposed to enhance detection efficiency and accuracy. These include expanding the dataset using data augmentation, incorporating attention mechanism modules to strengthen feature representation, designing a center-matching strategy to balance the number of positive samples, designing a shortest-distance label assignment strategy to eliminate ambiguous samples, and further improving detection accuracy through result-processing algorithms.

Finally, ablation and comparative experiments were conducted on an NVIDIA RTX 3080 GPU platform. The ablation experiments demonstrated that the proposed optimization methods effectively improved model detection accuracy. The optimized model achieved a 2.4% increase in CDF1 score and a 9.0% increase in mF1 score compared to the original YOLOv5 model on the cork disc dataset. The final optimized model reached a CDF1 score of 95.1%, a processing speed of 178.5 FPS, and an mAP of 81.5%. Compared to mainstream algorithms like Faster RCNN, RetinaNet, and CenterNet, the improved algorithm achieved the best detection performance on the cork disc dataset while maintaining high processing speed. Future work will deploy this algorithm on embedded development platforms and explore its application in other optical inspection fields.

Author Contributions: Conceptualization, L.Q.; methodology, G.C. and L.Q.; software, X.Z.; formal analysis, K.L.; investigation, K.L.; data curation, G.C. and X.Z.; writing—original draft preparation,

G.C.; writing—review and editing, L.Q.; funding acquisition, L.Q. All authors have read and agreed to the published version of the manuscript.

Funding: This research was funded by the Wuhu Science and Technology Project (No. 2022jc07) and the Graduate Student Innovation and Entrepreneurship Practice Project of Anhui Provincial Department of Education (No. 2022cxcysj058, 2023cxcysj043).

Institutional Review Board Statement: The study did not require ethical approval.

Informed Consent Statement: The study did not involve humans.

Data Availability Statement: Data are contained within the article.

Conflicts of Interest: Author Guohao Chen was employed by the Wuhan Mingke rail Transit Equipment Co., Ltd. The remaining authors declare that the research was conducted in the absence of any commercial or financial relationships that could be construed as a potential conflict of interest.

References

1. Díaz-Maroto, M.C.; López-Viñas, M.; Loarce, L.; Sanza, M.d.; Nevares, I.; Alañón, M.E.; Pérez-Coello, M.S. Quality control of natural cork stoppers by image analysis and oxygen transmission rate. *Holzforschung* **2022**, *76*, 863–873. [CrossRef]
2. Chang, S.H.; Han, G.H.; Valerde, J.M.; Griswold, N.C.; Duque-Carrillo, J.F.; Sanchez-Sinencio, E. Cork quality classification system using a unified image processing and fuzzy-neural network methodology. *IEEE Trans. Neural Netw.* **1997**, *8*, 964–974. [PubMed]
3. Gonzalez-Adrados, J.R.; Lopes, F.; Pereira, H. Quality grading of cork planks with classification models based on defect characterization. *Holz als Roh-und Werkstoff* **2000**, *58*, 39–45. [CrossRef]
4. Costa, A.; Pereira, H. Quality characterization of wine cork stoppers using computer vision. *J. Int. Sci. Vigne Vin.* **2005**, *39*, 209–218. [CrossRef]
5. Costa, A.; Pereira, H. Decision rules for computer vision quality classification of wine natural cork stoppers. *Am. J. Enol. Vitic.* **2006**, *57*, 210–219. [CrossRef]
6. Costa, A.; Pereira, H. Computer vision applied to cork stoppers inspection. In *Cork Oak Woodlands and Cork Industry: Present, Past and Future*; Zapata, S., Ed.; Museu del Surode Palafrugell Publications: Barcelona, Spain, 2009.
7. Georgieva, A.; Jordanov, I. Intelligent visual recognition and classification of cork tiles with neural networks. *IEEE Trans. Neural Netw.* **2009**, *20*, 675–685. [CrossRef]
8. Paniagua, B.; Vega-Rodríguez, M.A.; Gomez-Pulido, J.A.; Sanchez-Perez, J.M. Improving the industrial classification of cork stoppers by using image processing and Neuro-Fuzzy computing. *J. Intell. Manuf.* **2010**, *21*, 745–760. [CrossRef]
9. Oliveira, V.; Knapic, S.; Pereira, H. Classification modeling based on surface porosity for the grading of natural cork stoppers for quality wines. *Food Bioprod. Process.* **2015**, *93*, 69–76. [CrossRef]
10. Furtado, I.; Oliveira, A.S.; Amaro, F.; Lopes, P.; Cabral, M.; Bastos, M.d.L.; de Pinho, P.G.; Pinto, J. Volatile profile of cork as a tool for classification of natural cork stoppers. *Talanta* **2021**, *223*, 121698. [CrossRef] [PubMed]
11. Tang, H.; Zhu, H.; Fei, L.; Wang, T.; Cao, Y.; Xie, C. Low-Illumination Image Enhancement Based on Deep Learning Techniques: A Brief Review. *Photonics* **2023**, *10*, 198. [CrossRef]
12. Guan, J.; Li, J.; Yang, X.; Chen, X.; Xi, J. Defect detection method for specular surfaces based on deflectometry and deep learning. *Opt. Eng.* **2022**, *61*, 061407. [CrossRef]
13. Wongtanawijit, R.; Khaorapapong, T. Rubber tapping line detection in near-range images via customized YOLO and U-Net branches with parallel aggregation heads convolutional neural network. *Neural Comput. Appl.* **2022**, *34*, 20611–20627. [CrossRef]
14. He, Y.; Song, K.; Meng, Q.; Yan, Y. An end-to-end steel surface defect detection approach via fusing multiple hierarchical features. *IEEE Trans. Instrum. Meas.* **2020**, *69*, 1493–1504. [CrossRef]
15. Sharma, M.; Lim, J.; Lee, H. The amalgamation of the object detection and semantic segmentation for steel surface defect detection. *Applied. Sci.* **2022**, *12*, 6004. [CrossRef]
16. Wang, W.Y.; Mi, C.F.; Wu, Z.H.; Lu, K.; Long, H.; Pan, B.; Li, D.; Zhang, J.; Chen, P.; Wang, B. A real-time steel surface defect detection approach with high accuracy. *IEEE Trans. Instrum. Meas.* **2022**, *71*, 1–10. [CrossRef]
17. Mei, S.; Wang, Y.; Wen, G. Automatic fabric defect detection with a multi-scale convolutional denoising autoencoder network model. *Sensors* **2018**, *18*, 1064. [CrossRef]
18. Ren, X.Y.; Lin, W.Y.; Yang, X.Q.; Yu, X.; Gao, H. Data augmentation in defect detection of sanitary ceramics in small and non-i.i.d datasets. *IEEE Trans. Neural Netw. Learn. Syst.* **2022**, *34*, 8669–8678. [CrossRef]
19. Krizhevsky, A.; Sutskever, I.; Hinton, G. Imagenet classification with deep convolutional neural networks. *Commun. ACM* **2017**, *60*, 84–90. [CrossRef]
20. Girshick, R.; Donahue, J.; Darrell, T.; Malik, J. Rich feature hierarchies for accurate object detection and semantic segmentation. *IEEE Comput. Soc.* **2014**, 580–587.
21. Girshick, R. Fast R-CNN. In Proceedings of the IEEE International Conference on Computer Vision, Santiago, Chile, 7–13 December 2015; pp. 1440–1448.

22. Ren, S.Q.; He, K.M.; Girshick, R.; Sun, J. Faster R-CNN: Towards real-time object detection with region proposal networks. *IEEE Trans. Pattern Anal. Mach. Intell.* **2017**, *39*, 1137–1149. [CrossRef]
23. He, K.M.; Zhang, X.Y.; Ren, S.Q.; Sun, J. Deep residual learning for image recognition. In Proceedings of the IEEE Conference on Computer Vision and Pattern Recognition, Las Vegas, NV, USA, 27–30 June 2016; pp. 770–778.
24. Liu, W.; Anguelov, D.; Erhan, D.; Szegedy, C.; Reed, S.; Fu, C.Y.; Berg, A.C. SSD: Single shot multiBox detector. In Proceedings of the European Conference on Computer Vision (ECCV), Amsterdam, The Netherlands, 11–14 October 2016; pp. 21–37.
25. Lin, T.Y.; Goyal, P.; Girshick, R.; He, K.M.; Dollar, P. Focal loss for dense object detection. *IEEE Trans. Pattern Anal. Mach. Intell.* **2020**, *42*, 318–327. [CrossRef]
26. Duank, W.; Bai, S.; Xie, L.; Qi, H.; Huang, Q.; Tian, Q. CenterNet++ for Object Detection. *IEEE Trans. Pattern Anal. Mach. Intell.* **2024**, *46*, 3509–3521. [CrossRef]
27. Liu, Z.; Mao, H.Z.; Wu, C.Y. A ConvNet for the 2020s. In Proceedings of the IEEE/CVF Conference on Computer Vision and Pattern Recognition, New Orleans, LA, USA, 18–24 June 2022; pp. 11976–11986.
28. Redmon, J.; Divvala, S.; Girshick, R.; Farhadi, A. You Only Look Once: Unified Real-Time Object Detection. In Proceedings of the IEEE Conference on Computer Vision and Pattern Recognition, Las Vegas, NV, USA, 26 June–1 July 2016; pp. 5571–5579.
29. Redmon, J.; Farhadi, A. YOLO9000: Better, faster, stronger. In Proceedings of the IEEE Conference on Computer Vision and Pattern Recognition, Honolulu, HI, USA, 21–26 July 2017; pp. 7263–7271.
30. Edmon, J.; Farhadi, A. YOLOv3: An incremental improvement. *arXiv* **2018**, arXiv:1804.02767.
31. Bochkovskiy, A.; Wang, C.Y.; Liao, H.M. Yolov4: Optimal speed and accuracy of object detection. *arXiv* **2020**, arXiv:2004.10934.
32. Maktoof, M.A.J.; Al Attar, I.T.A.; Ibraheem, I.N. Comparison YOLOv5 Family for Human Crowd Detection. *Int. J. Online Biomed. Eng.* **2023**, *19*, 94–108. [CrossRef]
33. Ge, Z.; Liu, S.T.; Wang, F.; Li, Z.; Sun, J. YOLOX: Exceeding YOLO series in 2021. *arXiv* **2021**, arXiv:2107.08430.
34. Li, C.Y.; Li, L.L.; Jiang, H.L.; Weng, K.; Geng, Y.; Li, L.; Ke, Z.; Li, Q.; Cheng, M.; Nie, W.; et al. YOLOv6: A Single-Stage Object Detection Framework for Industrial Applications. *arXiv* **2022**, arXiv:2209.02976.
35. Wang, C.Y.; Bochkovskiy, A.; Liao, H.M. YOLOv7: Trainable bag-of-freebies sets new state-of-the-art for real-time object detectors. *arXiv* **2022**, arXiv:2207.02696.
36. Goodfellow, I.; Pouget-Abadie, J.; Mirza, M.; Xu, B.; Warde-Farley, D.; Ozair, S.; Courville, A.; Bengio, Y. Generative adversarial nets. *Commun. ACM* **2020**, *63*, 139–144. [CrossRef]

MDPI AG
Grosspeteranlage 5
4052 Basel
Switzerland
Tel.: +41 61 683 77 34

Photonics Editorial Office
E-mail: photonics@mdpi.com
www.mdpi.com/journal/photonics

www.ingramcontent.com/pod-product-compliance
Lightning Source LLC
Chambersburg PA
CBHW062049220926
51850CB00010B/2712
* 9 7 8 3 7 2 5 8 6 9 6 4 0 *